YA Non Fiction
YA 523.18 L58e

Lew, Kristi. The expanding universe
9001049760

D1716511

SCIENCE FOUNDATIONS

The Expanding Universe

SCIENCE FOUNDATIONS

The Big Bang
Cell Theory
Electricity and Magnetism
Evolution
The Expanding Universe
The Genetic Code
Germ Theory
Gravity
Heredity
Kingdoms of Life
Light and Sound
Matter and Energy
Natural Selection
Planetary Motion
Plate Tectonics
Quantum Theory
Radioactivity
Vaccines

SCIENCE FOUNDATIONS

The Expanding Universe

KRISTI LEW

Science Foundations: The Expanding Universe
Copyright © 2011 by Infobase Learning

All rights reserved. No part of this book may be reproduced or utilized in any form or by any means, electronic or mechanical, including photocopying, recording, or by any information storage or retrieval systems, without permission in writing from the publisher. For information, contact:

Chelsea House
An imprint of Infobase Learning
132 West 31st Street
New York, NY 10001

Library of Congress Cataloging-in-Publication Data
Lew, Kristi.
 The expanding universe / Kristi Lew.
   p. cm. — (Science foundations)
 Includes bibliographical references and index.
 ISBN 978-1-60413-292-2 (hardcover)
 1. Expanding universe—Popular works. I. Title. II. Series.

QB991.E94L49 2011
523.1'8—dc22   2010026878

Chelsea House books are available at special discounts when purchased in bulk quantities for businesses, associations, institutions, or sales promotions. Please call our Special Sales Department in New York at (212) 967-8800 or (800) 322-8755.

You can find Chelsea House on the World Wide Web at
http://www.infobaselearning.com

Text design by Kerry Casey
Cover design by Alicia Post
Composition by EJB Publishing Services
Cover printed by Yurchak Printing, Landisville, Pa.
Book printed and bound by Yurchak Printing, Landisville, Pa.
Date printed: June 2011
Printed in the United States of America

10 9 8 7 6 5 4 3 2 1

This book is printed on acid-free paper.

All links and Web addresses were checked and verified to be correct at the time of publication. Because of the dynamic nature of the Web, some addresses and links may have changed since publication and may no longer be valid.

Contents

1. Matter, Energy, and the Fundamental Forces — 7
2. Our Place in Space — 21
3. A Short History of Astronomy — 33
4. Einstein's Universe — 48
5. Make Sense of the Expanding Universe — 56
6. The Big Bang — 66
7. The Future Universe — 77

Glossary — 89
Bibliography — 91
Further Resources — 94
Picture Credits — 96
Index — 97
About the Author — 102

Matter, Energy, and the Fundamental Forces

The universe encompasses everything that is all around us. It contains all the stars, planets, solar systems, and galaxies. It is all visible and invisible matter. And it contains all forms of energy.

People have been curious about the universe and all that it contains since the beginning of time. Some of the ideas that were first set forth, such as the idea that Earth is the center of our solar system, have been proven incorrect. However, others, such as the idea that our galaxy is just one of the billions of galaxies that populate the universe, are now regarded as fact. Some of the things that scientists today suspect are true about the universe may also stand the test of time. On the other hand, some ideas may one day be proven to be untrue.

Scientists now believe that **matter** started to form in the first few microseconds after the universe was born. Matter is anything that has mass and takes up space. Anything that you can touch—a desk, a sandwich, a person—is made up of matter.

ATOMS

All matter is made up of smaller particles called **atoms**. Atoms, in turn, are made up of even smaller particles—**protons**, **neutrons**, and

electrons. Protons, neutrons, and electrons are called **subatomic particles** because they are smaller than an atom. By combining in different numbers, these subatomic particles form atoms of specific **elements**. Elements are chemical substances that cannot be broken down by ordinary chemical means such as heating them or passing electricity through them. Hydrogen and helium are elements. Gold is an element, too. Elements are made up of only one kind of atom.

All of the atoms of a particular element have the same number of protons. Protons are positively charged subatomic particles that are found in the **nucleus**, or central core, of every atom. For example, an atom that has one proton is a hydrogen atom. An atom with two protons is a helium atom. Gold atoms have 79 protons. An element's atomic number is equal to the number of protons in the atom's nucleus. All the known elements are listed in a chart called the periodic table of elements in order of increasing atomic number. The periodic table is a tool used by chemists to organize the elements according to the properties that they exhibit.

The protons in an atom's nucleus give the nucleus a positive charge. However, atoms are electrically neutral. In order for an atom to have a net neutral charge, the positively charged protons must be balanced by an equal number of negatively charged particles. These particles are the electrons. For example, the element helium has an atomic number of two. Therefore, every helium atom has two protons. In order for the atom to be electrically neutral, it must also have two electrons. Electrons move around the nucleus of the atom in energy levels.

Helium atoms also have two neutrons. Neutrons do not have a charge. They are electrically neutral, but they do have mass. Protons and neutrons have the same mass—1.66×10^{-27} kilograms, or 1 atomic mass unit (amu). All of the atomic masses listed on the periodic table, which are referred to as the mass number, are listed in atomic mass units. Electrons also have mass, but their mass is so tiny compared to the masses of the protons and neutrons that they do not have much of an influence on the mass number. According to the periodic table, the mass of a helium atom is 4.0 amu. Because each proton has a mass of 1 amu and there are two protons in every helium atom, each helium atom must also have two neutrons (2 protons = 2 amu + 2 neutrons = 2 amu, which adds up to a mass of 4 amu for a helium atom).

Neutrons are found in the nucleus of most, but not all, atoms. Hydrogen, for example, has an atomic number of 1. Hydrogen atoms, therefore, have one proton. The mass of most hydrogen atoms is 1 amu. This mass is the same as the mass of one proton, so most hydrogen atoms do not have any neutrons. However, not all hydrogen atoms are exactly alike; some of them do have neutrons. In fact, there are three naturally occurring types of hydrogen atoms. The most common type of hydrogen atom is ordinary hydrogen (also called protium) which has a mass of 1 amu. This type of hydrogen atom accounts for 99.985% of all the hydrogen that is found. Another type of hydrogen, however, has a mass of 2 amu. This type of hydrogen is called heavy hydrogen, or deuterium. Like ordinary hydrogen, deuterium also has one proton (otherwise, it would be an entirely different element). However, because deuterium has a mass of 2 amu, it also has one neutron. The third type of hydrogen is called tritium. Tritium has a mass of 3 amu. It has two neutrons. Atoms of the same element that have a different number of neutrons and, therefore, a different mass are called **isotopes**.

FUNDAMENTAL PARTICLES

Protons, neutrons, and electrons are not the only subatomic particles. In fact, protons and neutrons are not even the smallest subatomic particles. Both protons and neutrons are made up of **quarks**. At this time, scientists do not believe that quarks can be broken down into smaller particles. Because they cannot be broken down (as far as we know), quarks are called **fundamental particles**. Scientists have identified six types of quarks so far. Physicists refer to the different types of quarks as "flavors," and each flavor has its own name. The six flavors of quarks are "up," "down," "charm," "strange," "bottom," and "top." The six flavors are usually paired up as follows: up and down, charm and strange, and bottom and top.

Each flavor has its own particular mass (this is one of the ways that physicists can tell which flavor is which). The up and down quarks are the lightest. These are the quarks that make up protons and neutrons. Protons and neutrons are both made up of three quarks, but they each have a different combination of flavors. Protons, for example, are made up of two up quarks and one down

10 THE EXPANDING UNIVERSE

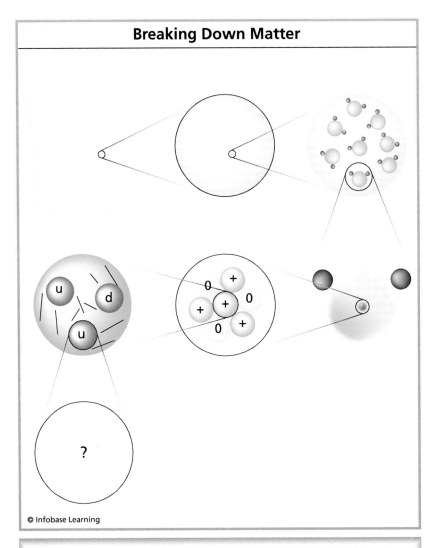

Figure 1.1 Zooming in on a glass of water, you can imagine the individual water molecules made up of one oxygen atom (green) and two hydrogen atoms (purple). When you zoom in on a molecule, you can see that the atoms have a nucleus made up of protons and neutrons (*middle center*). Those protons and neutrons are each made up of three quarks (*middle left*). Scientists do not yet know what the quarks are made up of.

quark. Neutrons, on the other hand, are composed of one up quark and two down quarks. The heavier quarks (charm, strange, top, and bottom) do not exist in nature, but scientists have been able to produce them in the laboratory.

Matter, Energy, and the Fundamental Forces 11

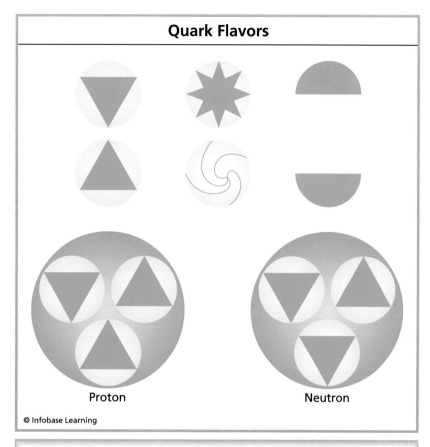

Figure 1.2 Quarks come in six flavors: up, down, strange, charmed, top, and bottom. A proton is made of two up quarks and one down quark, while a neutron is made of two down quarks and one up quark.

This explains what makes up protons and neutrons, but what makes up electrons? At this time, scientists believe that the electron is not made up of any other particles. They are a flavor of another type of fundamental particle called a lepton. Like quarks, leptons cannot be broken down into smaller particles. Leptons also come in six flavors, and the electron is one of those flavors. The other lepton flavors are muon, tau, electron neutrino, muon neutrino, and tau neutrino. Like quarks, each lepton has its own mass. Electrons are the lightest of the lepton flavors.

Physicists have found about 200 types of particles in their search for more fundamental particles. Most have turned out not to be fundamental, but scientists are still looking. In the meantime, they use a theory called the standard model to explain how everything in the world is made up of six different types of quarks and six different flavors of leptons, and to explain how these particles are held together by other types of particles that are called force carriers.

FUNDAMENTAL FORCES

Quarks and leptons are the building blocks of matter, but the fundamental forces of nature are what hold quarks and leptons together. There are four fundamental forces—the **electromagnetic force**, the **strong force**, the **weak force**, and **gravity**.

Like the poles of magnets, charges repel or attract one another. Like charges repel just as two north poles of two magnets would repel each other if they were brought close to one another. Also, just as a north pole and a south pole of two magnets would attract one another, so do unlike charges. In other words, a positively charged proton and a negatively charged electron would attract each other. Two positively charged protons, however, would repel one another. The force that causes charged particles to act in this manner is called the electromagnetic force. The electromagnetic force is what holds whole atoms together. It also holds everything made up of those atoms—including Earth and humans—together.

The electromagnetic force, which includes light, is carried by what physicists call a force carrier. The force carrier that carries the electromagnetic force is called a photon. Photons travel at the speed of light, and they do not have mass. Photons can be absorbed by charged particles like protons and electrons. Charged particles can also produce photons. However, photons cannot be absorbed or produced by an uncharged particle like a neutrino. Different photons have different amounts of energy. The differing energy amounts span the entire electromagnetic spectrum, from radio waves to X-rays, including visible light.

The electromagnetic force seems to hold atoms together because of the attraction between protons and electrons, but it does not explain what holds an atom's nucleus together. Protons are charged particles, but neutrons are not. So, what holds these neutral particles in the nucleus together with the positively charged protons?

And why don't those protons that all have the same charge repel each other and blow the nucleus apart? The answer is that there is a different force, called the strong force or strong nuclear force, which holds the nucleus together. The particles that carry the strong force are called gluons (because they "glue" quarks together). The strong force is stronger than the electromagnetic force. Therefore, the strong force is able to overcome the repulsive electromagnetic force between protons.

One other fundamental force is at work at the atomic level, as well. This force is called the weak force. The weak force is responsible for the decay (or break down) of some quarks and leptons. The atoms that make up all of the matter around us are composed of the lightest flavor of quarks—the up and the down quarks—and the least massive lepton—the electron. In order to form up and down quarks and electrons, other and more massive quarks and leptons need to decay. When quarks or leptons decay into another type of quark or lepton, physicists say that they have changed flavors. All flavor changes are caused by the weak force. The weak force is also the force that causes some types of nuclear reactions, such as the radioactive decay of certain radioactive elements. Particles called weak bosons carry the weak force.

The last of the four fundamental forces of nature is gravity. Gravity works on a much larger scale than the electromagnetic, strong, and weak forces. Gravity is what keeps the Moon orbiting Earth and Earth orbiting the Sun. It is also the force that keeps our solar system and galaxy together in a cohesive system. However, gravity is one of nature's weakest forces. It is also the one that is the least understood by scientists. Because all the other forces are carried by some sort of particle, scientists predict that there is also an elementary particle that carries the gravitational force. However, this force carrier has not been found yet. Nevertheless, this theoretical force carrier has been given a name: a graviton.

ENERGY

The interaction between matter and energy controls everything in the universe. Matter can be changed from one form of matter into another by adding energy, for example. Matter can exist in one of

14 THE EXPANDING UNIVERSE

Antimatter

For every type of particle that scientists have found so far, there is a corresponding antimatter particle, or antiparticle. For example, the antiparticle of a quark is called an antiquark. The proton's antimatter particle is called an antiproton. Pretty easy so far, but the electron's antiparticle has a different type of name: a positron.

All of the antimatter particles have the same mass as their ordinary matter counterpart, but they have the opposite charge. When antimatter and matter meet, they completely destruct. What was once matter turns into pure energy. For example, when an electron and a positron meet, they annihilate each other and produce two gamma rays. (A gamma ray is a form of energy.)

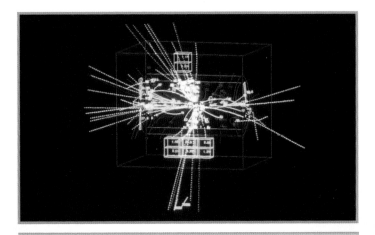

Figure 1.3 This computer model shows jets of particles created when matter and antimatter collide and annihilate in a particle accelerator at CERN, the European particle physics laboratory near Geneva. The model shows an event in which two quarks have collided and annihilated into pure energy, rematerializing as two new quarks which in turn decay into two high-energy jets (blue and purple tracks) of hadronic matter, such as pions. The display also shows the yellow and green tracks of lower energy particles produced from the event. The red outline shows the UA1 detector, which is the size of a three-story house, in which the collision occurred.

four main states: solid, liquid, gas, or plasma. It can move between these states, but not without adding or taking away energy. In fact, the main difference between these states of matter is how much energy they contain. Take ice, for example. Ice is water that is frozen into a solid form of matter. Ice can be changed into other forms of matter by adding energy, or heat, to it. Add a little bit of energy and the solid ice turns into liquid water. Add more energy to the liquid water and steam, a gas, will be formed. A gas contains more energy than a liquid. Plasma, a state of matter that can only exist at very high temperatures, contains even more energy. Heat is one form of energy, but there are others, as well, including light and movement (sometimes called mechanical energy).

Energy can change from one form to another, too. For example, a ball at the top of a hill contains a form of energy called potential energy. Potential energy is stored energy and can be changed into kinetic energy. Kinetic energy is the energy of motion. When the ball starts to roll down the hill, its potential energy is converted into kinetic energy.

All astronomical bodies, such as stars, planets, and comets, give off another form of energy, which is called electromagnetic energy, or electromagnetic radiation. Electromagnetic radiation is energy that travels in the form of electromagnetic waves. Unlike sound waves or waves in a body of water, electromagnetic waves do not need to travel through matter. In fact, they can travel through the vacuum of space. Gamma rays, X-rays, ultraviolet (UV) light, visible light, infrared (IR) light, microwave, and radio waves are all different types of electromagnetic radiation. Electromagnetic waves are categorized by their **wavelengths**. A wavelength is the distance from the top of one wave to the top of the next. Radio waves have the longest wavelength and gamma rays have the shortest.

The types and amounts of electromagnetic energy given off by an astronomical body depend on the energy that the body contains. Gamma rays are the most energetic forms of electromagnetic energy while radio waves contain the smallest amount of energy of the waves in the electromagnetic spectrum. Astronomers use different instruments to detect the different wavelengths of electromagnetic energy that are emitted by particular astronomical bodies. For example, a telescope can be used to detect visible light coming from a star, but because our eyes can only see a very small part of the electromagnetic spectrum, astronomers also use other instruments.

Radio telescopes, for example, can detect radio waves. Infrared telescopes and microwave telescopes are also used to study distant stars, planets, and galaxies.

DARK MATTER

Gravity is not the only thing in the universe that astronomers cannot quite explain. In 1933, Fritz Zwicky (1898–1974), an astronomer at the California Institute of Technology in Pasadena, noticed that a particular cluster (or group) of galaxies appeared to be moving faster than other galaxy clusters that had the same mass. The problem with the speed at which this cluster was traveling is that, if the mass of the cluster that Zwicky calculated was correct (and he was fairly sure that it was), then those galaxies should not be able to hold together. The cluster did not contain enough mass to create a gravitational pull strong enough to keep its galaxies from flying off into space.

Gamma Ray Bursts

One type of electromagnetic radiation observed by astronomers is called a gamma ray burst. Gamma ray bursts are caused by a burst of gamma ray photons and are very short lived. Gamma ray photons are the most energetic form of light and are also very bright—about a million trillion times brighter than the Sun. Scientists have observed gamma ray bursts emitted by supernovae. A **supernova** is the explosion of a massive star that takes place when the star dies.

Some of the gamma ray bursts seen by astronomers have occurred at the very edge of the observable universe. Because of the universe's size, it takes light a very long time to travel from the edge of the universe to Earth. Therefore, scientists believe that some of the gamma ray bursts that they are now observing came from stars that exploded just as Earth was forming—before the first microbes existed, before the oceans formed, and way before the first humans emerged.

Figure 1.4 Swiss astronomer Fritz Zwicky first theorized that galaxies must contain more matter than he could actually see in order to produce the gravity required. The asteroid 1803 Zwicky and a lunar crater are both named in his honor.

Zwicky proposed that maybe the galaxy cluster actually contained more mass than it appeared to have. In other words, he thought that the galaxies must contain more matter than he could actually see in order to produce the gravity required. Today, astronomers call this matter **dark matter**.

18   THE EXPANDING UNIVERSE

We cannot see dark matter because it emits no light. In fact, dark matter emits no radiation at all. So, if scientists cannot see dark matter, how do they know that it is there? The answer is that scientists can see the effects of dark matter. Dark matter is detected through a phenomenon called **gravitational lensing**. Gravitational lensing occurs when an object is so massive that it can bend light. Astronomers have detected parts of the universe that they believe contain a lot of dark matter because light coming from distant galaxies is bent by these areas as it travels to Earth, where it is observed by the astronomer's instruments. In some cases, the dark matter bends the light from these galaxies so much that the astronomers see double and quadruple images of faraway objects. Dark matter seems to be clumped in certain parts of the universe, primarily in the halos of galaxies.

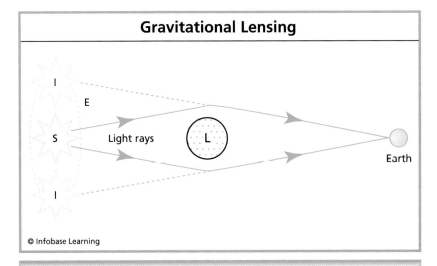

Figure 1.5   A gravitational lens is a massive astronomical object, such as a star, galaxy, or dark object, that is located between a distant luminous object, such as a galaxy, and Earth. It bends light rays emanating from the distant object on their way to Earth. In this way, the light source viewed on Earth might appear distorted. This illustration represents two light rays in an ideally symmetric lensing event, wherein the source (S) would appear to an observer as a ring (E; called an Einstein ring) with the lensing (bending) object (L) at the ring's center. To an observer on Earth, the two light rays appear as two images (I) of the source, which form the Einstein ring.

Scientists are still in the process of determining exactly what makes up dark matter. So far, they have ruled out the idea that dark matter is made up of clouds of dust. There would need to be way too much dust to account for dark matter's gravitational pull. Scientists have also determined that dark matter is not any type of hydrogen; nor is it an ordinary black hole in its active state because active black holes give off X-rays when objects enter them, and dark matter does not emit any kind of electromagnetic radiation. It is possible that some dark matter is actually ordinary matter that does not give off enough light for astronomers to detect it—objects such as planets, brown dwarfs, and dead stars, for example. Scientists believe that these objects may account for up to about 50% of dark matter. They think that the rest of the dark matter out there may be made up of weakly interacting massive particles (WIMPs). Scientists believe that WIMPs may be some type of subatomic particle. So far, the theory is that WIMPs have mass, but they interact weakly with other particles. Otherwise, they would bump into ordinary matter all the time and, therefore, would be much easier to detect.

Even though Zwicky proposed the existence of dark matter in 1933, scientists still do not know much about it. Currently, scientists believe that about 21% of the universe is made up of dark matter and that it exerts more than 80% of the gravitational force in the universe. Only about 4% of the universe is made up of ordinary matter, and the rest is made up of **dark energy**.

The Difference Between Dark Matter and Antimatter

Dark matter and antimatter are not the same things. When an ordinary matter particle, such as a proton, and its antimatter particle, an antiproton, meet, they destroy each other and produce a burst of energy. However, when an ordinary matter particle meets a dark matter particle, all they do is give each other a little gravitational tug.

DARK ENERGY

Like dark matter, dark energy cannot be seen. However, unlike dark matter, dark energy does not seem to be gathered in certain places in the universe. Instead, it is spread out all over the universe in a relatively uniform manner. Still, dark energy is even more important in determining the future of the universe than dark matter is. That is because dark energy seems to be speeding up the universe's expansion. Dark energy acts, essentially, in the opposite way that gravity acts. Scientists think that, on a cosmic scale, dark energy is constantly pushing superclusters away from each other. It is also pushing superclusters of galaxies away from each other at faster and faster speeds. Within the galaxies themselves, however, gravity is still the stronger force. Gravity keeps the stars together in their galaxies, which is why stars are not scattered throughout the universe.

Scientists have not known about dark energy all that long—only since the late 1990s. Thus, like dark matter research, research into exactly what dark energy is and what it does is ongoing. In fact, some scientists do not believe that dark energy exists at all and claim that we just do not know enough about the force of gravity to explain it thoroughly.

Our Place in Space

Scientists who study the universe, its history, and all it contains are called cosmologists or astronomers. Today, astronomers know quite a lot about the universe. For example, they know that it is made up of gas, dust, stars, planets, and galaxies. They also know that the universe is unimaginably large.

THE SCALE OF THE UNIVERSE

People used to believe that the universe consisted of only the objects that they could see in the night sky. Today, we know that the universe is much larger. In fact, the visible edge of the universe is about 10 trillion (10,000,000,000,000) times further away from Earth than Neptune, the most distant planet in our solar system.

A space this large is much too big to measure in terms of miles or kilometers (km). The Moon, for example, is 238,855 miles (384,400 km) away from Earth. Hundreds of thousands of miles is not such a large number, but to get to Venus, the closest planet to Earth, you would have to travel about 25,717,910 miles (41,388,950 km). This is still not an unimaginably large number, but it is big enough to become cumbersome when making calculations.

To help keep the numbers that correspond to the vast space taken up by our universe easier to calculate, astronomers developed a unit of measurement called an astronomical unit (AU). An AU is equal to the average distance between Earth and the Sun, or about 93 million miles (150 million km). Using astronomical units, the distance between Earth and the Moon becomes 0.00257 AU. And the distance between Earth and Venus becomes a more manageable 0.277 AU.

Another way to measure distances in the universe is in **light-years**. In space, light travels at a constant speed of approximately 186,000 miles (300,000 km) per second. A light-year is the distance that light travels in one year, about 63,240 AU or 5.8 trillion miles (9.5 trillion km). Expressed in terms of light-years, the distance from Earth to the Moon is 1.282 light-seconds. In other words, on average, it takes 1.282 seconds for light to travel from the Moon to Earth. Pluto, the dwarf planet on the edge of our solar system, is about 5.5 light-hours away from Earth. The next closest star to our solar system, Proxima Centauri, is 4.2 light-years away. Because the light from Proxima Centauri takes 4.2 years to reach Earth, astronomers observing the star do not actually see what Proxima Centauri looks like today. Instead, they are seeing what the star was like 4.2 years ago. The next closest galaxy, the Andromeda Galaxy, is 2.9 million light-years away. The light that astronomers observe from this galaxy today was emitted before humans appeared on Earth. As you can imagine, space is a big place.

THE STRUCTURE OF THE UNIVERSE

Earth is part of a **galaxy** called the Milky Way galaxy. A galaxy is a group of stars, dust, and gas that is held together by gravity. Our galaxy is part of a group of galaxies that scientists call the Local Group. The universe is made up of billions of galaxies that are all strung together in a structure that resembles a gigantic spider web.

Galaxies

Our Sun is only one of the billions of stars that populate the Milky Way galaxy. Galaxies come in a variety of shapes. The Milky Way is a spiral galaxy, the most common type. Of the galaxies that scientists have observed so far, 77% are spiral galaxies. This type of galaxy has

Our Place in Space 23

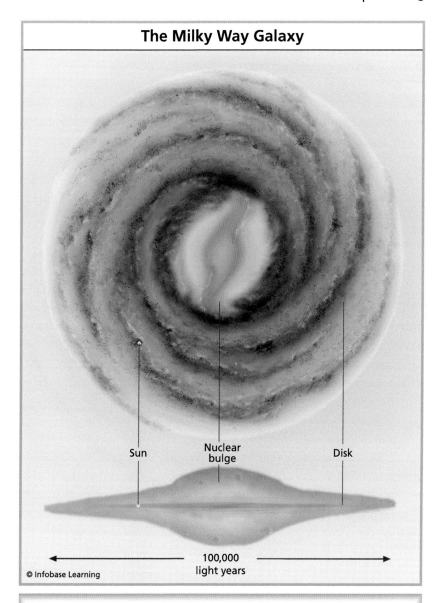

Figure 2.1 The Milky Way is a spiral galaxy consisting of more than 400 billion stars as well as gas and dust arranged into the components: the halo, a spherical pattern that contains the oldest stars in the galaxy; the nuclear bulge and galactic (rotational) center; and the disk, which contains the majority of the stars, including the Sun, and gas and dust.

a bulge in the center with long arms that spiral out from it. Some spiral galaxies have a bright line running through their centers. Scientists call these galaxies "barred spiral galaxies."

Galaxies can also be elliptical. The shape of an elliptical galaxy can vary from an almost completely circular shape to a long, thin, cigar shape. Another type of galaxy, called a Lenticular galaxy, looks like a spiral galaxy with no arms. In fact, scientists sometimes call Lenticular galaxies "armless spiral galaxies." Like spiral galaxies, Lenticular galaxies sometimes have a bright line running through their centers and are called barred Lenticular galaxies. Most galaxies fall into one of these three main categories, but about 3% do not. These galaxies are called irregular.

Earth lies on one of the Milky Way's spiral arms and is about 28,000 light-years away from the center of the galaxy. The Milky Way is a part of a group of galaxies called the Local Group. For light to get from one side of the Local Group to the other would take about 7 million light-years. Our Local Group contains about 40 galaxies. These 40 galaxies are also a part of a larger group called the Local Supercluster. The Local Supercluster contains roughly 4,000 galaxies and is about 60 million light-years across. Scientists believe that there are as many as 10 million superclusters in the universe. Between superclusters, the universe is mostly empty space.

Scientists believe that the oldest galaxies and **quasars** formed about 800 million years after the birth of the universe. Quasars, which stands for "quasi-stellar radio sources," are some of the oldest and brightest objects in the universe. In fact, some quasars can shine with the intensity of 1,000 large galaxies. They were first discovered in the 1960s by Maarten Schmidt (1929–) and Jesse Greenstein (1909–2002). Schmidt and Greenstein found that quasars are located very far from Earth, which also made them incredibly old. Today, scientists believe that quasars are the centers of very distant galaxies. They also believe that the quasars that they are looking at today are at least "second generation" quasars. In other words, earlier quasars existed before them. Someday, by using better technology than is currently available, scientists may be able to observe even older ones.

Stars

There are about 200 billion stars in our galaxy. Our Sun is just one of them, and is the one that is closest to Earth. The Sun's diameter is

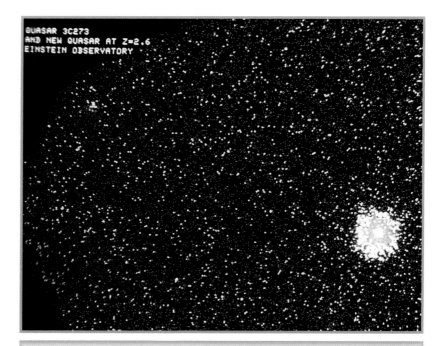

Figure 2.2 Quasars are mysterious, bright, starlike objects seemingly located at the edge of the visible universe. This image, taken by the High Energy Astronomy Observatory 2/Einstein Observatory, shows Quasar 3C 273. Scientists estimate that it is 10 billion light years away.

about 863,706 miles (1.39 million km), making it the largest object in our solar system, by far.

Our Sun is classified as a yellow dwarf star. Stars in this classification are relatively small and have a surface temperature between 9,080°F and 10,340°F (5,300 Kelvin and 6,000 K). The surface temperature of the Sun is approximately 9,944°F (5,780 K). The Sun's core is much hotter, closer to 27 million°F (15 million K).

The Sun, like most other stars, is a main-sequence star. Main-sequence stars make up the central band on the Hertzsprung-Russell diagram. The Hertzsprung-Russell diagram is a graph on which astronomers plot a star's color, or temperature, versus its luminosity, or brightness. In general, the hotter main-sequence stars are the brighter ones.

Main-sequence stars, including the Sun, are fueled by **nuclear fusion** reactions. In a nuclear fusion reaction, the nuclei of smaller atoms fuse to make the nuclei of larger atoms. In this process,

26 THE EXPANDING UNIVERSE

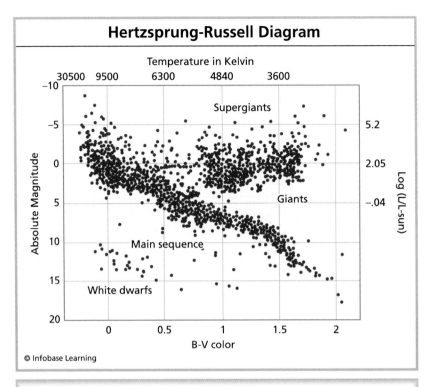

Figure 2.3 In a Hertzsprung-Russell diagram, each star is represented by a dot. The position of each dot corresponds to the star's temperature and absolute magnitude (brightness).

nuclear fusion reactions also produce large amounts of energy. Currently, the Sun is made up of approximately 70% hydrogen and 28% helium by mass. The other 2% is made up of other elements. These percentages change, however, as hydrogen is converted into helium during nuclear fusion. At the Sun's core, every second approximately 700 million tons (635 million metric tons) of hydrogen nuclei are combined to form 695 million tons (630 million metric tons) of helium and 5 million tons (4.5 million metric tons) of energy.

Scientists believe that our Sun formed approximately 4.3 billion years ago. It is estimated that it will continue to burn for another 7 billion years or so at which point it will run out of fuel. As long as a star has enough fuel to carry out nuclear fusion, the outward force of the radiation produced in this process balances out the inward force of the star's gravity. When a star runs out of fuel, it starts to collapse.

Scientists believe that when our Sun runs out of fuel, it will also collapse and become a white dwarf star as all small stars (up to about 8 times as massive as our Sun) typically do. White dwarf stars are white, dim stars that are below the main-sequence stars on the Hertzsprung-Russell diagram. The more massive the star, the more it will contract (because of the greater gravitational force) and the smaller it becomes when it dies. On Earth, a teaspoon of material from a white dwarf star would weigh more than five and a half tons (4.5 metric tons). The cores of these dying stars shrink to about the size of Earth and cool. Eventually, the star cools enough to become a cold, dark black dwarf.

Unlike our Sun, scientists believe that the first stars in the universe were made up of only hydrogen and helium. When massive, fast-burning stars use up their nuclear fuel, they tend to collapse suddenly in a spectacular explosion called a supernova. In 1987, a supernova lit up the night sky when a star called Sanduleak -69° 202 exploded in a

Black Holes

In the late 1700s, the English geologist John Michell (1724–1793) came to the conclusion that, theoretically, an object could have so much gravity that nothing, not even light, would be able to escape it. These objects would have to be very, very massive and dense to create this kind of gravity. Michell called these objects "dark stars." In the 1960s, however, physicist John Wheeler (1911–2008) coined the term "black hole" to describe these objects. Their density causes their gravitational force to be exceptionally strong, so strong, in fact, that not even light can escape the gravitational hold of a black hole.

Depending on a star's mass and make up, it may become a white dwarf or a neutron star when it dies. However, if the star is massive enough, it may collapse in on itself and form a black hole instead. The "edge" of a black hole, or the area where light just fails to escape, is called the event horizon. The event horizon separates a black hole from the rest of space, and any event happening inside it cannot be seen.

nearby galaxy called the Large Magellenic Cloud. (The Large Magellenic Cloud is 160,000 light-years away from Earth.) Originally, Sanduleak -69° 202 was about 20 times more massive than our Sun. This supernova was the first one to be visible from Earth in about 400 years. The blast from a supernova creates a massive amount of energy that can fuse lighter elements into heavier elements. The explosion scatters these elements throughout the universe. Astronomers believe that the heavier elements found in our Sun are the remnants of a previous generation of stars going supernova.

In some cases, if a star was massive enough, a supernova produces a neutron star. Neutron stars are very small, super-dense stars. On Earth, a teaspoon of material from a neutron star would weigh over 100 million tons (90.7 million metric tons). Neutron stars that rotate rapidly are called pulsars (short for pulsating stars). Pulsars give off pulses of microwave or X-ray radiation like a lighthouse sends out pulses of light. These pulses can be detected by instruments on Earth. Over the last 30 years, scientists have found over 300 pulsars scattered across the universe. The timing of the "on" and "off" flashes of pulsars can be so accurate, they can be used to measure time very accurately. Scientists believe that the explosion of Sanduleak -69° 202 caused its core to contract into a rotating neutron star, but they have not been able to confirm this yet.

Pulsars are not the only super-dense objects in our universe that can give off X-ray radiation. Black holes do, too. In the early 1970s, the Uhuru X-ray satellite detected signals coming from the constellation Cygnus. They named the system Cygnus X-1 because it was the first X-ray source found in that constellation. It was also the first suspected black hole. Cygnus X-1 is about 8,000 light-years away from Earth and it is a binary star system, meaning that it is made up of two stars. One of the stars is a blue supergiant that is about 400,000 times brighter and 30 times more massive than our Sun. Supergiants are the largest known stars, and they are very rare. Some of these stars can be as large as our solar system. When supergiants die, they supernova and become black holes.

Scientists estimate that the other star in Cygnus X-1 was 5 to 10 times more massive than our Sun. Scientists believe that when this star died, it collapsed into a super-dense core that formed a black hole. Scientists are able to detect the invisible black hole as it sucks

Our Place in Space 29

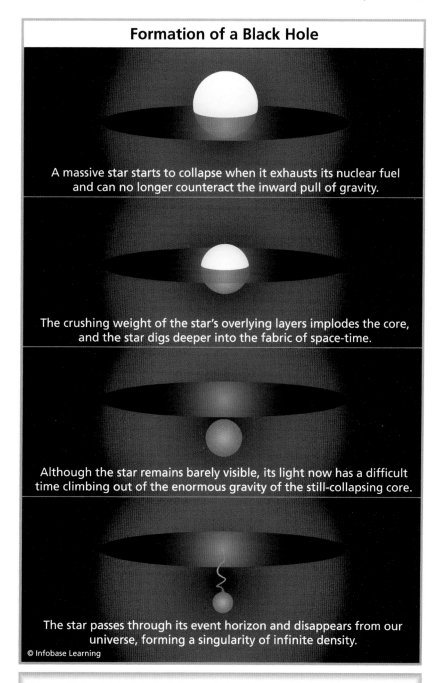

Figure 2.4 A black hole is formed when a star collapses in on itself due to gravitational force.

Exoplanets

In 2004, Hawaii's Gemini North telescope and the telescopes at the W.M. Keck Observatory captured the first infrared pictures of an exoplanet, a planet outside of our solar system. Four years later, in November 2008, the Hubble Telescope took the very first visible light pictures of an exoplanet called Fomalhaut b. The exoplanets that have been photographed so far are large, gaseous planets like Jupiter. Still, scientists hope that one day, they might be able to find and photograph a very Earth-like planet. Some scientists believe they will need Earth-based instruments that are 1,000 to 10,000 times stronger than the ones they have today to capture good, crisp, visible light photographs of these exoplanets.

Figure 2.5   This image, taken by Hawaii's Gemini North telescope, shows an exoplanet that is a composite of J-, H- and K-band near-infrared images.

gas off its companion blue supergiant. This gas swirls around the edge of the black hole and emits short, powerful bursts of X-rays.

Cygnus X-1 is not the only binary star system in our galaxy. In fact, scientists believe that nearly 70% of the stars in the Milky Way exist in star systems that contain more than one star. Polaris (also called the North Star, Northern star, or pole star) is part of a binary star system, too.

Polaris is a blue-green **Cepheid** variable star. Cepheid stars are giant stars that change in brightness and size on a regular basis. As the size of the star increases, its brightness decreases and then when the reverse occurs, the star gets smaller and brighter again. Polaris's brightness and size change about every four days.

Our Solar System

At this time, scientists estimate that the universe is between 13.5 and 13.9 billion years old. They believe that the Milky Way formed from a cloud of dust and gas, called a **nebula**, shortly after the universe was formed. Our solar system, however, is considerably younger. According to current observations, scientists believe that the solar system formed when gravity drew the dust and gas in the nebula closer and closer together, increasing temperatures significantly. About 5 billion years ago, the temperature was hot enough to spark nuclear reactions and the Sun was formed. The material in the cloud that was left over after the formation of the Sun then cooled and condensed into solid particles. Over the next hundreds of thousands of years, these solid particles clumped together under the force of gravity. Some of these clumps became quite large. These big balls of heavier, rockier material stayed close to the Sun and became the terrestrial planets of Mercury, Venus, Earth, and Mars. The gas giants—Jupiter, Saturn, Uranus, and Neptune—formed in the outer solar system from the left over gas and dust.

Planets are not the only bodies that travel around the Sun. Scientists have found more than 100 moons orbiting the planets in our solar system, too. Some planets, like Earth, only have one satellite (the Moon) while others, like Jupiter, have many more.

There are thousands of other rocky bodies, called asteroids, moving around the Sun, as well. The largest of these is about 621 miles (1,000 km) across. Most of these asteroids can be found in

Figure 2.6 The world's most famous comet, Halley's Comet, is shown here as photographed by a wide-angle camera at Lowell Observatory in Arizona in 1910. A streak across the comet is a meteor trail.

what astronomers call the asteroid belt between the planets of Mars and Jupiter.

Comets also occasionally plow through the solar system on their way around the Sun. These rocky-icy objects produce gases and dust that trail behind them in a long tail as they whiz through the solar system. So far, scientists have identified at least 150 comets that appear regularly in our solar system.

A Short History of Astronomy

Humans have always been interested in the sky above them and what it contains. Cosmologists study stars, galaxies, solar systems, planets, comets, asteroids, quark-gluon plasma, and black holes in an attempt to explain how the universe came to be and how it might eventually end. To do their jobs, today's scientists build on the discoveries of the countless other scientists who studied the universe before them.

THE ANCIENTS

The first astronomers were the ancient Babylonians, who started studying the skies above around 1,000 B.C. Believing that their fates depended on the patterns and positions of the stars, these ancient people studied the stars closely, even going so far as to map them. The 12 signs of the Zodiac that astrologists use today are a part of their legacy.

Eratosthenes (276–196 B.C.), a Greek astronomer who lived in Alexandria during the third century B.C., was the first scientist to measure the circumference of Earth. He noticed that while the Sun appeared directly overhead at noon in Syene (now Aswan), Egypt, it appeared slightly (about 7°) south when seen from Alexandria. Assuming that all the Sun's rays were perpendicular to Earth when they hit the planet's curved surface and knowing the distance between Alexandria and Syene, Eratosthenes used geometry to calculate the circumference of Earth. His calculations showed that the planet must be about 25,000 miles around (40,234 km). Considering that modern techniques have shown the circumference of Earth to be 24,902 miles (40,076 km) at the equator, Eratosthenes was astonishingly close. (Because Earth is an oblate ellipsoid, or a slightly flattened sphere, its circumference is a bit shorter at the poles: 24,860 miles).

A GEOCENTRIC VIEW

Another Greek astronomer, Claudius Ptolemaeus (circa A.D. 100– A.D. 170), known as Ptolemy, believed that Earth was at the center of the universe. In his geocentric view, the Sun and the other five planets known during his lifetime traveled around Earth in perfectly circular paths at constant speeds. He believed that they orbited Earth with Mercury and Venus in front of the Sun and Mars, Jupiter, and Saturn behind it. Using his system, Ptolemy was able to predict where any of these planets would be at a given time.

Ptolemy also believed that the stars were just a backdrop on which the planets and the Sun moved around Earth. His work included a star catalog that contained 48 constellations. We still use the names Ptolemy gave these constellations today.

Ptolemy's geocentric idea of the universe persisted for 1,400 years, but his system had a hard time accounting for certain observations; for instance, that certain planets sometimes traveled east-to-west across the sky instead of their normal west-to-east path. This path of travel is called **retrograde motion**.

Ptolemy explained this motion by adding concentric circles to his system that showed each planet moving around a smaller circle which, in turn, moved around a larger circle that centered on Earth. The final model was very complicated with many layers of nested circles that were called epicycles.

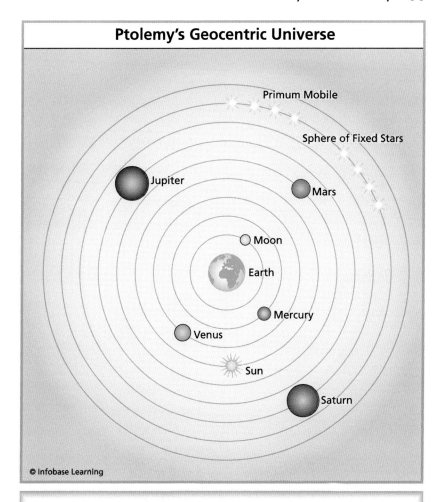

Figure 3.1 Ptolemy believed that Earth was at the center of the universe, circled by the Moon, Mercury, Venus, the Sun, Mars, Jupiter, and Saturn.

THE SUN AS THE CENTER

In 1530, Nicolaus Copernicus (1473–1543), a Polish astronomer, proposed that if the Sun was placed in the center of the solar system rather than Earth, Ptolemy's system would be much less complicated. (Copernicus was not the first astronomer to propose placing the

36 THE EXPANDING UNIVERSE

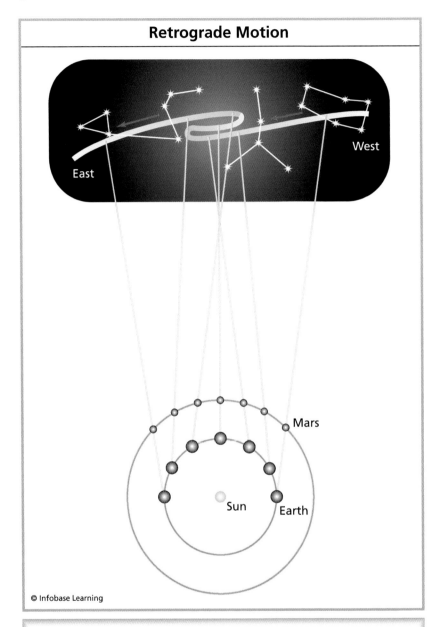

Figure 3.2 A celestial object's inclination indicates whether an object's orbit is in direct or retrograde (opposite) motion.

Sun in the center of the solar system. The Greek astronomer Aristarchus of Samos suggested the same thing around 250 b.c., but no one paid any attention to him.)

Copernicus's model was still based on each planet following a perfectly circular path, but it did not need to include epicycles to explain retrograde motion. This omission made Copernicus's system much simpler than Ptolemy's. In Copernicus's system, retrograde motion was a natural occurrence of Earth's and the other planet's motion around the Sun. Retrograde motion occurs because planets closer to the Sun move around their orbits faster than planets farther away. As Earth goes around the Sun faster than Mars, for example, it overtakes, or "laps" Mars. From the perspective of someone observing Mars on Earth, Mars seems to slow down and reverse direction—just as a runner on the outside lane of a track appears to move backward relative to the crowd from a faster runner's perspective. On the background of stationary-appearing distant stars, Mars seems to make the same motion.

Copernicus wrote a book explaining his findings, but he did not publish it right away because he knew that his ideas went against the teachings of the Catholic Church. The Church would consider his idea of the Sun being at the center of the solar system instead of Earth to be heresy, a crime often punishable by death during that time period. Therefore, Copernicus waited more than 10 years to publish his ideas. The book was finally published in 1543 when Copernicus was already on his deathbed. Historians are not sure if he lived to see his book in print, but a preface was added (some say without Copernicus's knowledge) saying that his theory was intended only to make the mathematical calculations of the positions of the planets easier. The preface went on to add that Copernicus's model was not intended to be an actual model of what was really happening in the solar system.

REFINING THE COPERNICAN MODEL

Near the end of the sixteenth century, the Danish astronomer Tycho Brahe (1546–1601) made many contributions to astronomy. In 1572, he observed a supernova in the constellation Cassiopeia. Brahe became a respected astronomer after he carefully recorded his observations and published his findings regarding this new "star." He also performed a 20-year study of the motion of many of the planets, especially Mars. From his observations, Brahe did not believe that Ptolemy's geocentric version of the universe was correct. However,

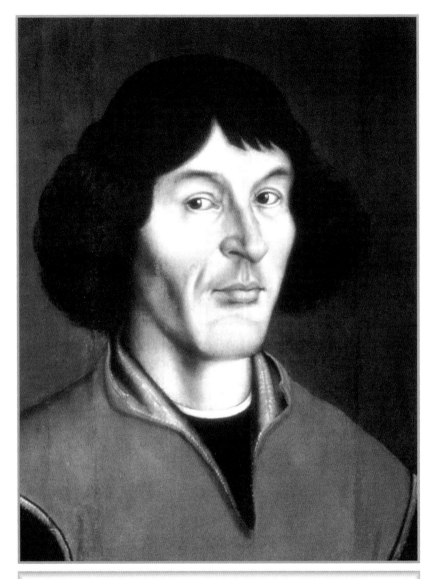

Figure 3.3 Nicolaus Copernicus was one of the first to formulate a heliocentric model of the universe, which displaced Earth as the center.

he did not believe Copernicus's heliocentric version, either. Like Ptolemy, Brahe believed that Earth was the center of the universe and that the Sun and the Moon revolved around Earth. Unlike Ptolemy, however, Brahe believed that all the other planets revolved around the Sun and not around Earth.

Brahe also compiled one of the most accurate catalogs of the stars seen up to that point—it listed more than 1,000 stars that he had observed. This was not the most comprehensive list of stars available then, but it was the most accurate. What is more amazing is that he did all this work with the naked eye—the first telescope was not invented until nine years after his death.

None of Brahe's observations were published during his lifetime, but in 1600, he hired an assistant, Johannes Kepler (1571–1630), who used Brahe's work to further the science of astronomy. After Brahe's death, Kepler continued to study his observations and tried to match Brahe's position of Mars with all sorts of circular paths. He was unsuccessful. Finally, Kepler proposed that perhaps the path of the planets were ellipses, or slightly flattened circles, instead of perfect ones. An elliptical path would almost perfectly fit the motion of Mars as observed by Brahe.

The planets' elliptical path around the Sun became the first of Kepler's three laws of planetary motion. Kepler's second law stated that when a planet is at its perihelion, the point in its orbit that is closest to the Sun, it moves faster than when it is at its aphelion, the point in its orbit when it is the farthest away. His third law related the distance of a planet from the Sun to how quickly it completed its orbit. Mercury, the planet that is closest to the Sun, for example, takes only 88 days to complete its orbit while Neptune takes 60,190 days.

Copernicus' theory of a heliocentric universe, Brahe's observations, and Kepler's calculations did not really come into favor, however, until Galileo Galilei (1564–1642), an Italian mathematician and astronomer, made his observations with the newly invented telescope. Many people believe that Galileo invented the telescope, but that distinction actually belongs to Dutch eyeglass maker Hans Lippershey (1570–1619). The first record of a small telescope made by Lippershey appeared in 1608. Galileo heard about the invention that made distant objects appear closer then they actually are in the following year, 1609. Within a month, Galileo had duplicated this invention.

Using his knowledge of mathematics, Galileo was able to improve on the existing telescope. In 1610, although his telescope magnified objects only about 30 times, Galileo was able to view and describe the craters on the Moon. Up until this point, people thought the Moon to be absolutely smooth. Galileo went on to find the four largest moons of Jupiter using his telescope. He also discovered that the phases of Venus mimicked the phases of the Moon, and made a study of Saturn and sunspots. To ensure that the general public could read about his discoveries, Galileo communicated his results in Italian rather than the more academic Latin.

The Scientific Method

As Johannes Kepler was developing his laws of planetary motion, he kept meticulous records that documented how he came to his conclusions. Today, we would say Kepler was following the scientific method. The scientific method has several steps:

1. **Ask a question.** Kepler, for example, wanted to know why Brahe's observations did not fit a perfect circle.
2. **Do background research or make observations.** Kepler used Brahe's 20 year study of the planets' motions.
3. **Form a hypothesis (educated guess).** Kepler eventually decided that maybe the planets did not orbit the Sun in a perfect circle.
4. **Test with an experiment.** Kepler tried many different patterns of circles and other planetary models before drawing a conclusion.
5. **Analyze data and draw a conclusion.** After careful analysis, Kepler found that the only shape that would fit all of Brahe's observations of planetary motion was an ellipse.
6. **Report results.** In 1609, Kepler published his first two laws in *Astronomia Nova* (New Astronomy).

A Short History of Astronomy 41

Figure 3.4 Galileo is pictured discussing his astronomical theories with a monk. In 1616, the Catholic Church condemned his heliocentric view of the universe, noting that it was "false and contrary to scripture."

Galileo's observations convinced him that Copernicus was correct—the Sun, not Earth, was the center of the universe. He used his observations and the scientific method similar to the one used by scientists today to back up his theory. Like Copernicus, Galileo quickly ran into trouble with the Catholic Church when he started publishing papers describing his Sun-centered universe. In 1615, the Inquisition (the legal body of the Church) accused Galileo of heresy, or opposing the Church's teachings. Galileo was cleared of the charges. However, he was told that he should no longer state his belief that Earth orbited the Sun. From there, Galileo went back to work. Over time, however, he became more and more convinced that Copernicus was correct that Earth and all of the other planets moved around the Sun. In 1632, he published a book called *Dialogue*

42 THE EXPANDING UNIVERSE

Concerning the Two Chief World Systems. Again, Galileo stated that he agreed with the Copernican heliocentric model of the universe. And again, the Inquisition called Galileo to Rome and charged him as a heretic. This time Galileo was found guilty and was sentenced

Stephen Hawking

On January 8, 1942—exactly 300 years after Galileo Galilei died—Stephen Hawking (1942–) was born in Oxford, England. Hawking is one of today's greatest theoretical physicists. His contributions to the field of cosmology have been numerous. In his third year at Oxford University, however, Hawking noticed that he seemed to be having trouble with muscular coordination. At first, he paid no attention to the problem. The following year, however, his father noticed his uncoordinated movements and took him to a doctor.

Shortly after his twenty-first birthday, Hawking was diagnosed with an incurable motor neuron disease called amyotrophic lateral sclerosis (ALS), or Lou Gehrig's disease. This disease causes deterioration of motor cells (the ones that control the way we move, speak, breathe, and swallow) in the brain and spinal cord. The disease does not normally affect the intellect, memory, or the senses. Hawking's doctors did not expect him to live more than a few years. In most ALS patients, the progression of the disease is rapid. Life expectancy is usually only two to five years after the onset of symptoms. However, Hawking has beaten the odds and has already survived for over four decades with the disease.

After receiving both his diagnosis and his doctorate in cosmology, Hawking has spent his career studying the basic laws that govern our universe. An even larger achievement than his theoretical physics work is possibly Hawking's scientific writing for non-scientists. His first popular science book, *A Brief History of Time*, published in 1988, became an international bestseller and spent more than four years on the London *Sunday Times* bestseller list.

to life in prison. Because of his age and poor health, however, the Church allowed Galileo to serve out his life sentence in his home. Galileo spent the last decade of his life under house arrest and died on January 8, 1642.

Figure 3.5 Stephen Hawking, a professor of mathematics at the University of Cambridge, is shown delivering a speech at George Washington University's Morton Auditorium in Washington, D.C., on April 21, 2008.

44 THE EXPANDING UNIVERSE

Figure 3.6 This is a faithful replica of the first reflecting telescope, which was made by Sir Isaac Newton. To use the telescope, the viewer looks through an eyepiece on the side of the tube. The light path is reflected off both the primary mirror and a flat secondary one near the top of the tube.

The type of telescopes developed by Lippershey and Galileo are called refracting or refractor telescopes. The lenses in these telescopes are the same types of lenses used today in binoculars and telephoto camera lenses. Refracting lenses can gather more light than the human eye and can, therefore, present a brighter and more magnified image. There were, however, some drawbacks to the refractor telescope made by Lippershey and Galileo. They often produced unwanted rings around bright objects, such as stars, for example, and there was a limit to their magnification. One way to solve these problems was to make the telescopes longer. Where one of Galileo's telescopes might have been 5 or 6 feet (1.5 or 1.8 m) long, other scientists were making refractor telescopes that were 15 to 20 feet (4.6 to 6.1 m) long. Some of the telescopes were as long as 100 feet (30.5 m). This made the telescopes quite hard to carry and maneuver in tight spaces.

In 1668, Isaac Newton (1642–1727) improved on Galileo's design by replacing the lens with a mirror. Today, these telescopes are called Newtonian telescopes. Newtonian telescopes are a type of reflecting telescope. Reflecting telescopes use mirrors to focus the light from distant objects. This same basic technology is still used in nearly all of today's large telescopes.

The reflecting telescope was just one of Isaac Newton's many accomplishments. Newton made many discoveries that affected the worlds of astronomy, mathematics, and physics. He invented calculus, discovered three very important laws of motion, and developed the law of universal gravitation. Newton's law of universal gravitation states that every object in the universe attracts every other object in the universe due to its gravity. Gravity, acting at a distance, causes two bodies to accelerate toward one another. The strength of this force depends on how massive and how close together the objects are. He found that gravitational force is proportional to the product of the masses of the objects and inversely proportional to the distance between them. Therefore, larger masses create a larger gravitational pull than smaller masses. And objects that are close together exert a stronger gravitational pull on one another than objects that are farther away.

Using his law of universal gravitation and the newly invented calculus, Newton was able to show that, mathematically, Kepler's laws were correct. Newton explained that the gravitational force of the Sun keeps planets and other solar system bodies in orbit around it. He also explained that the elliptical paths of the planets were due to gravity.

In the late 1800s, however, scientists discovered a problem with Newton's theory of gravitation. As Mercury orbits the Sun, its path shifts such that its closest point to the Sun, its perihelion, moves forward 5 arc minutes every century creating what is known as the

46 THE EXPANDING UNIVERSE

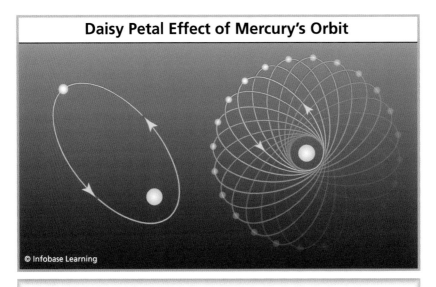

Figure 3.7 As Mercury orbits the Sun, its path shifts slightly forward with every pass, thereby creating a pattern called the daisy petal effect. Newton's calculations accounted for some of this shift. In 1916, Einstein's theory of relativity explained the rest.

"daisy petal effect." Newton's calculations accounted for some of this shift, but they were off by 42 arc seconds (an arc second is one-sixtieth of an arc minute. An arc minute is one-sixtieth of a degree).

However, Newton's laws remained the foundation of cosmological models until Einstein appeared in the early twentieth century with another way of looking at the universe. In his mid-twenties, Newton was named the Lucasian Professor of Mathematics at Cambridge. The same post was held by Stephen Hawking between 1979 and 2009.

THE MILKY WAY

It was not until the beginning of the eighteenth century that astronomers began to suspect that the Sun was only one of a group of stars. At that time, most people still believed that if the Sun was indeed a part of a galaxy, then the Sun must be the center of it. In 1918, however, American astronomer Harlow Shapley (1885–1972) was able to determine the shape and size of the Milky Way and the Sun's place in it by studying the earlier work of Henrietta Leavitt (1868–1921). In 1912, Leavitt discovered that there was a connection between the brightness of a Cepheid variable star and the amount of time between its pulses. Leavitt found that the longer a Cepheid's pulsation period, the more luminous the star. Because the luminosity of a star is also related to its distance from Earth, a Cepheid's pulsation period can also tell astronomers how far away the star is. Using Leavitt's data, Shapley was able to determine that the Milky Way was much larger than anyone at the time suspected—around 100,000 light-years across. And that the Sun is not at the center of the galaxy. It is actually about 30,000 light-years from the center.

Einstein's Universe

At the beginning of the twentieth century, Albert Einstein (1879–1955) proposed his ideas of how matter, energy, space, time, and gravity interact with each other. Einstein was born in Ulm, Germany. Growing up in Munich, Einstein was not a great student. While working in a patent office in Zurich, Switzerland, however, Einstein spent most of his spare time studying physics. In 1905, the self-taught physicist published three papers that would turn the physics world on its collective ears.

One of these papers described mathematically the random movements of tiny particles. Another reported the way certain metals emit electrons when exposed to light, a phenomenon called the photoelectric effect. In 1921, Einstein was awarded his only Nobel Prize in physics for the discovery of this effect. Einstein's last paper written that year described his theory of special relativity.

EINSTEIN'S SPECIAL THEORY OF RELATIVITY

Einstein's special theory of relativity describes how matter and light travel at high speeds. The theory states that the speed of light in a vacuum is always the same: 186,411 miles per second (300,000 km/s). It also states that no wave or particle can travel faster than the

Einstein's Universe

Figure 4.1 Albert Einstein is pictured in 1947.

speed of light. These ideas have a profound effect on how high-speed particles are perceived by an observer. For example, objects traveling near the speed of light will look shorter than they actually are

50 THE EXPANDING UNIVERSE

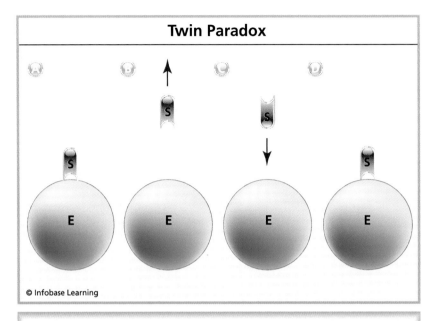

Figure 4.2 In an example of twin paradox, when an earthbound twin greets his twin sister upon her return from a trip into space and back, she is younger than he is. Their different rates of aging can be attributed to the fact that during the trip, the earthbound twin remains in a single inertial reference frame, while the traveling twin finds herself in four consecutive inertial reference frames and undergoes an acceleration from each frame to the next. (a) She and her spaceship S are in her first inertial reference frame, that of Earth E. Then she accelerates to (b) her second inertial reference frame, which is in constant-velocity motion away from Earth. Another acceleration turns her around and puts her in (c) her third inertial reference frame, moving toward Earth at constant velocity. Her third acceleration—a deceleration as viewed from Earth—takes her to rest on Earth and into (d) her fourth inertial reference frame.

to a stationary observer. However, if the observer is traveling at the same speed as the object, the object's length would appear normal. In other words, the appearance of the object is relative to the position and speed of the observer.

In his special theory of relativity, Einstein also puts forth the idea that, mathematically, time and space are not independent of one another. Instead, Einstein said they are connected in a single fabric called space-time. In other words, stars, galaxies, cars, and people all travel through a combination of the two. Masses in the universe, such as planets, for example, bend the fabric of space-time. In turn, this bend in space-time influences how the mass moves and also influences its properties.

Because space and time are intrinsically tied together, Einstein proposed that the passage of time depends on where a body is in the universe and how fast it is traveling. For example, as an object travels closer and closer to the speed of light, the object seems to slow down when seen by an outside, stationary observer. At the speed of light,

The World's Most Famous Equation

Einstein's special theory of relativity also showed how the energy released from a radioactive element is related to its mass. Over time, the nuclei of radioactive elements break down to form lighter elements and energy. The amount of energy released is equal to the amount of mass lost multiplied by the speed of light squared, or as it is more simply stated in the following, famous equation:

$$E = mc^2$$

Therefore, energy can be created from mass. Because the speed of light is high, this equation means that even a small amount of matter can be converted into a large amount of energy. Inside stars, for example, lighter elements are fused into heavier ones. In this process, a tiny amount of mass is lost. This mass is converted into energy that keeps the stars burning.

Not only can mass be turned into energy, but energy can also be turned into mass. Scientists believe that there was a huge amount of energy at the beginning of the universe. In turn, they believe that collisions between high-energy gamma rays could have produced all of the matter and antimatter that exist in the universe today. Some of these particles would have encountered their antiparticle, annihilated each other, and returned energy back to the universe. Others would remain electrons, positrons, and quarks that would eventually form the protons and neutrons needed to form atoms and, in turn, larger chunks of matter such as rocks, planets, and humans.

to that observer, time would appear to stop. However, if the observer was moving at the same speed as the object, time would appear to pass normally. Therefore, time is also relative to the observer's speed and position.

These ideas led Einstein to the conclusion that the faster an object moves through space, the slower it moves through time. Scientists have tested this using an atomic clock traveling in a high-speed jet. When compared to a stationary atomic clock on the ground, the atomic clock in the jet ticks slower.

Einstein's theory also predicted how speed would affect the mass of an object. The faster an object moves through space, the more massive it becomes and the thinner it appears in the direction of its motion to an observer watching it. However, these effects on the appearance of objects and time only become apparent at very high speeds (approximately 10% the speed of light or about 18,611 miles per second, or 30,000 km/s). This is why a person whizzing past you on a bicycle does not look any thinner, but biker going fast enough would appear much skinnier as he or she drew closer to the speed of light.

GENERAL THEORY OF RELATIVITY

Einstein's special theory of relativity only applies to bodies that are not in a gravitational field. Eleven years after Einstein developed his special theory of relativity, he incorporated gravity into his equations. The general theory of relativity unifies Einstein's special theory of relativity and Newton's theory of universal gravitation. In this theory, Einstein incorporated how gravity affects space-time. Einstein's general theory of relativity states that space-time is an elastic substance that can contract, expand, or warp.

Einstein explained that smaller bodies, such as Earth, are not attracted to larger objects, like the Sun, by an invisible gravitational force acting at a distance, as Newton thought. Instead, large, massive objects like the Sun warp the space-time fabric like a bowling ball would warp a rubber sheet. If a smaller object such as a tennis or golf ball, for example, were laid on the rubber sheet, the smaller ball would roll toward the bowling ball because of the "dent" it makes in the rubber sheet. In other words, gravity is a warping of the space-time fabric created by the presence of matter.

The general theory of relativity was able to explain the "daisy petal effect" of Mercury's orbit around the Sun that Newton's theory could not. Because Mercury is the closest planet to the Sun, its orbit shifts a little bit every time it passes close to the star. This shift is caused by the Sun's warping of the space-time fabric.

TESTING GENERAL RELATIVITY

In 1919, three years after Einstein published his general theory of relativity, while on an island off the west coast of Africa, Sir Arthur Eddington (1882–1944) performed the first experimental test of Einstein's new theory. During a solar eclipse, Eddington took pictures of light rays emitted by a star cluster named Hyades. This star cluster lies near the Sun, about 151 light-years from Earth. If Einstein's general theory of relativity was correct, the light from these stars should appear to bend as the mass of the Sun warps the fabric of space-time. During the eclipse, Eddington also sent another group of scientists to Brazil to take pictures of the same group of stars. When the scientists compared their photographs, they did indeed see a shift in light coming from the star cluster just as Einstein predicted.

Today, astronomers call this phenomenon gravitational lensing because massive objects can focus and enlarge distant objects, just like a telescope lens does. However, sometimes very dense objects like black holes, galaxies, or galaxy clusters that act as gravitational lenses can also bend light from objects behind them to create images such as arcs, rings, or even multiple images of the object.

Astronomers can use gravitational lensing to "see" dark matter in galaxy clusters. This works a little bit like seeing an image through a glass full of water. Even though water is transparent, objects behind the glass of water appear distorted because the water bends light rays. Likewise, dark matter is also transparent, but its gravity bends light rays to reveal its presence. By analyzing the way light rays are bent in certain areas of the universe, astronomers can determine the distribution of dark matter.

IS TIME TRAVEL POSSIBLE?

Einstein believed that there could be a "bridge" that connected distant regions of space-time. These bridges were originally called

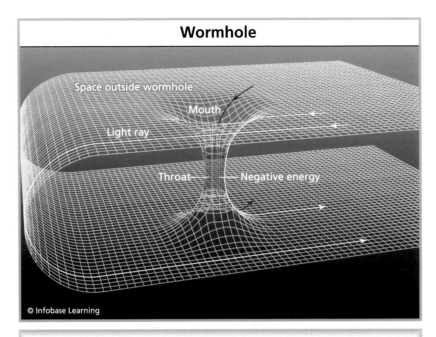

Figure 4.3 A wormhole is a hypothetical mathematical feature of space-time that is like a shortcut through time. This image shows an analogy to a wormhole.

Einstein-Rosen bridges in honor of co-discoverers Einstein and Nathan Rosen (1909–1995) who first proposed the idea in 1935. Today, they are called wormholes.

Indeed, Einstein's general theory of relativity states that time is not a constant. Instead, it moves more like a river, speeding up and slowing down, as it moves around massive bodies in the universe. Therefore, one second on Earth is not the same as one second on Mars. Kurt Gödel (1906–1978), a scientist at Princeton University, used Einstein's theory to reason that, theoretically, there could be whirlpools in the fabric of space-time. In these whirlpools, time would essentially curve back onto itself. Gödel's solution to Einstein's theory would allow for time travel.

In 1963, Roy Kerr (1934–), a mathematician from New Zealand, also used Einstein's theory to describe how a rotating black hole would not necessarily collapse into a single point (a singularity) as previously thought. It could turn into a spinning ring of neutrons that would not collapse in on itself because it was spinning so fast. Kerr reasoned that the spinning ring of neutrons could create enough centrifugal force to counteract gravity and create a path from one part of space, and, therefore, one part of time to another.

Before anyone can hop into a wormhole and travel though time, however, there are some significant problems that would need to be overcome first. The first problem is energy. Just as cars need gasoline, time machines would need some kind of fuel and they would need a lot of it. Until recently, physicists believed that if wormholes did exist, they would only last for a split-second before collapsing into a singularity. However, more recent calculations have shown that, with enough energy, it might be possible to keep the wormhole from disappearing. However, the amount of energy required could only come from a couple of places. If scientists could figure out how to harness the energy of a star, for example, they might be able to do it. Another possibility is to find a source of negative energy. Until recently, scientists did not think negative energy existed, but a tiny amount of it has been produced experimentally. However, large amounts of negative energy are incredibly hard to produce—at least for now.

Energy is only one of the problems facing a time-traveler. The wormhole's stability could be a problem, too. It is very possible that Kerr's rotating black hole would become unstable if a person or an object were to enter it. Disturbances on the atomic level could cause a wormhole to collapse. Without a theory that unifies quantum mechanics and gravity, however, it is very hard for physicists to predict the exact stability of a wormhole. So, for now, time travel is not a possibility—at least not until physicists completely understand and can overcome the problems of adequate energy and wormhole stability.

Today, Einstein's theories still make up much of the foundation of theoretical physics. However, there were some implications of his general theory of relativity that not even Einstein was prepared to consider.

Making Sense of the Expanding Universe

According to Issac Newton's theories, the universe never changed. In his thinking, the universe was the same during his lifetime as it always had been. In other words, the universe was static. However, when Einstein developed his general theory of relativity, he ran into a problem. His equations predicted that the universe should either be contracting or expanding. Because Einstein, and most other scientists at the time, agreed that the universe was static, Einstein thought he had made a mistake in developing his mathematical formulas. In 1922, Russian mathematician and physicist Alexander Friedmann (1888–1925), also mathematically proved that, if Einstein's general theory of relativity was correct, space could not remain stationary. Still entrenched in the idea of a static universe, however, Einstein would vigorously oppose this idea.

In fact, Einstein was so convinced that the universe could not be expanding or contracting that he added a number he called the cosmological constant to his general relativity equation. The cosmological constant acted as a repulsive force to counteract the pull of gravity. By making this repulsive gravitational constant exactly equal to the attractive gravity of matter, Einstein was able to construct a completely static model of the universe. He later called the addition of the cosmological constant "the greatest blunder of my life."

STELLAR DISTANCE

To determine whether or not the universe is expanding, contracting, or standing still, astronomers needed to know exactly how other stars and galaxies were moving relative to Earth. First, they had to find out how far away the stars were. Astronomers use several methods to determine the distance between a star and Earth. One method is called parallax. Parallax is the apparent motion of a close-up object as compared to the background. To see an example of parallax, try putting a finger close to your face, close one eye, and look at your finger from the other eye. Now switch eyes. Your finger will appear to shift positions compared to objects in the background. That appearance of motion is parallax. Astronomers use parallax to measure the distance to stars that are relatively close to Earth. To use parallax, scientists use very sensitive equipment to measure the position of a nearby star relative to more distant stars in the background. Six months later, when Earth is on the opposite side of its orbit around the Sun, researchers repeat the measurement. Even for the closest star, the difference in position will be tiny—less than an arc second. Astronomers have been measuring parallaxes for hundreds of years and they have accurate measurements of a few thousand stars. This method works best for stars that are no more than a few hundred light-years away. For stars that are farther away, their parallax is too small to be detected by the equipment that scientists are currently using.

For stars that are beyond a few hundred light-years away, scientists have to rely on other methods for measuring distance. One method is to use a star's brightness, or luminosity. The color of a star and its actual brightness are related to each other. Because they are related, scientists can determine a star's luminosity from its color. By comparing the star's actual luminosity with the apparent brightness of the starlight once it has reached Earth, they can determine the star's distance.

In astronomy, the Doppler effect can be applied to the light given off by stars. This can tell scientists if the stars are stationary, moving toward Earth, or moving away. If stars are moving toward Earth, their light waves are compressed. Therefore, the wavelength of the light is shorter, shifting the color of the light toward the blue end of the visible light spectrum. Scientists call this effect blueshift. On the other hand, if stars are moving away from Earth, scientists would expect the wavelength of the light to be stretched. Longer wavelengths shift the light toward the red end of the spectrum. This effect is known as redshift.

THE EXPANDING UNIVERSE

More than a decade after Einstein proposed his general theory of relativity, American astronomer Edwin Hubble (1889–1953) discovered that nearly every galaxy in the universe exhibited a redshift. If distant galaxies exhibit redshift, they must be moving away from Earth.

Hubble determined a galaxy's distance from Earth by observing the pulsating stars called Cepheids. Almost every galaxy has at least one of these stars. Cepheids pulsate in a particular rhythm. Henrietta Leavitt (1868–1921) was the first scientist to realize that the brightness of

Doppler Effect

Have you ever noticed that a siren on a fire truck, police car, or ambulance sounds different depending on whether it is coming toward you or moving away from you? This phenomenon is called the Doppler effect. This effect was first discovered by Austrian mathematician and physicist Christian Doppler (1803–1853). A siren sounds different because its pitch is higher as it comes toward you and becomes lower as it races away. This change in pitch is caused by a shift in the frequency of the sound wave. As the vehicle moves toward an observer, the sound waves being emitted are compressed resulting in a higher frequency and, therefore, a higher pitch. As the vehicle moves away from the observer, the sound waves are stretched out. The longer wavelength translates into a lower frequency and pitch. The Doppler effect holds true for any wave-emitting object, no matter whether the waves being emitted are sound, light, or radio waves.

Making Sense of the Expanding Universe 59

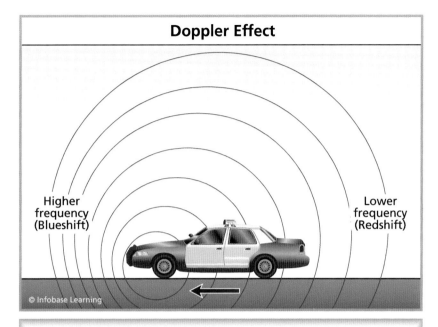

Figure 5.1 The Doppler effect is the change in wavelength (and, hence, frequency) of a wave motion as a result of the relative motion of the source and the observer. For a source moving away from the observer, the observed wavelength is longer than it would be if the source and observer had no relative motion along the line joining them. This change to longer wavelengths (toward the red end of the visible spectrum for visible wavelengths) is called a redshift. Conversely, if the source is moving toward the observer, there is a change to shorter wavelengths, called a blueshift.

these stars varies with time in a periodic fashion. While working at the Harvard Observatory, Leavitt observed the presence of many Cepheids in the Milky Way. She published her observations in 1912. In 1922, a decade after Leavitt's observations, Hubble used the most powerful telescope available at that time, the 100-inch (254-centimeter) Hooker telescope at the Mount Wilson Observatory, to observe Cepheids in other galaxies. By comparing the brightness of Cepheids in other galaxies with those in the Milky Way that pulsated at the same rate, Hubble was able to estimate the distance to the other galaxy. Today, the basic method used by Hubble to determine the distance to another galaxy is still used. However, better technology makes it easier to calculate this distance than it was during Hubble's lifetime.

60 THE EXPANDING UNIVERSE

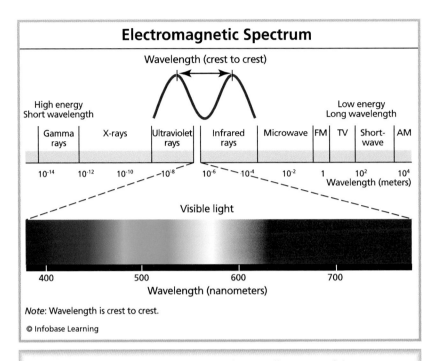

Figure 5.2 Visable light makes up only a small portion of the electromagnetic spectrum. Higher energy corresponds to a shorter wavelength and higher frequency, while lower energy corresponds to longer wavelengths and lower frequency.

Hubble also used Leavitt's Cepheid data and his own observations to determine how fast the galaxies were moving away. To do this, he measured the color of the Cepheids. All stars give off a particular pattern of colors depending on the makeup of their atmospheres. Specifically, scientists use the fact that all stars have hot hydrogen gas in their atmosphere to determine the star's rate of motion. If a star is stationary relative to Earth, the pattern of colors given off by it should be the same pattern of colors given off by hot hydrogen gas produced in a laboratory on Earth. However, if a star is moving away, the time it takes for the crest of each light wave to reach Earth is longer, which produces a longer wavelength. This means that the color of the star has redshifted, and the faster a star is moving away from Earth, the greater the redshift. With further observations, Hubble also found that the more distant the galaxy, the greater its redshift. Therefore, Hubble's conclusion was that not only are galaxies moving away from Earth and each other, but that they do so at a speed proportional to their distance. In other words, the more distant the galaxy, the faster it is traveling away from Earth.

The faster an object is moving, the more its wavelength is stretched. Therefore, the faster an object is moving away from Earth, the more redshifted its spectrum. In fact, the wavelengths of more distant, dimmer galaxies are so redshifted that they are much harder to detect. This is because the wavelength of the light emitted is stretched so much, it passes out of the visible spectrum and into the infrared part of the electromagnetic spectrum. Astronomers call this phenomenon cosmological dimming.

When Hubble graphed the speed of the galaxies versus their distance, he found that the speed was directly proportional to the galaxy's distance from Earth. In other words, the universe was stretching in a uniform manner. The slope of the straight line plotted by Hubble is called the Hubble parameter and it is a measure of the rate of expansion of the universe. This rate is called Hubble's constant. One of the main projects of the Hubble Space Telescope is to determine the value of this constant. Hubble's findings are now called Hubble's law. Together, these observations convinced Hubble and other scientists that our universe is not static—it is expanding.

WILL IT CONTINUE TO EXPAND?

The temperature, density, and composition of the universe all depend on the rate of expansion. Therefore, scientists are very interested in whether or not the universe will continue to expand, and, if it does, will the expansion slow down or will it speed up?

One theory proposes that the universe will keep expanding outward—faster and faster—and that galaxies will continue to get farther and farther away from each other. This is called the accelerating universe theory.

Scientists have found that for the first 7 billion years or so of the universe's existence, the rate of the universe's expansion slowed. They believe that this was because gravitational attraction was overwhelming the repulsive force that would have allowed the universe to continue to expand. Galaxies increased in size and stars formed quickly during this time. Then some scientists believe that around 6 to 7 billion years ago, as galaxies drew further apart and a repulsive force took over, the expansion of the universe sped up again at an increasingly accelerated rate. At this point, the rate of star formation fell about 90%. The timing of the onset of this expansion is disputed by other scientists, however. Some scientists believe that the accelerated expansion began only recently. At this time, scientists believe that the universe has always been expanding. However, the expansion seems to be speeding up. They are not exactly sure what is causing this acceleration, but the prevailing theory, at the moment, is that it may be caused by dark energy.

FATE OF THE UNIVERSE

Whether or not the universe continues to expand will depend on the total mass (ordinary matter plus dark matter) of the universe and its interaction with dark energy. Scientists have developed three models of the universe based on the amount of matter it contains. If, in the future, it turns out that there is enough mass in the universe to cause galaxies to stop their outward rush and reverse direction, the universe will be what scientists call a closed universe. If there is not enough mass to stop the expansion, it will be an open universe. And if the universe continues to expand, but slows to the point where the expansion is basically zero, it will be a flat universe. Based on these models, scientists have constructed three main scenarios for the ultimate fate of the universe—the big crunch, the big chill, and the big rip.

The big crunch theory explains what might happen if our universe is closed. In this scenario, expansion slows, eventually stops, and then reverses. The universe is pulled inward and collapses. Galaxies would get closer and closer. Eventually, they would condense down to one extraordinarily dense point called a singularity. However, the big crunch theory no longer matches the observations of astronomers. Therefore, most scientists have discarded this theory.

If our universe turns out to be open, scientists believe it will end in what is called the big chill. This theory is also called the eternal universe. In this scenario, galaxies would get farther and father apart. All stars would eventually run out of fuel and die. After the stars

Making Sense of the Expanding Universe 63

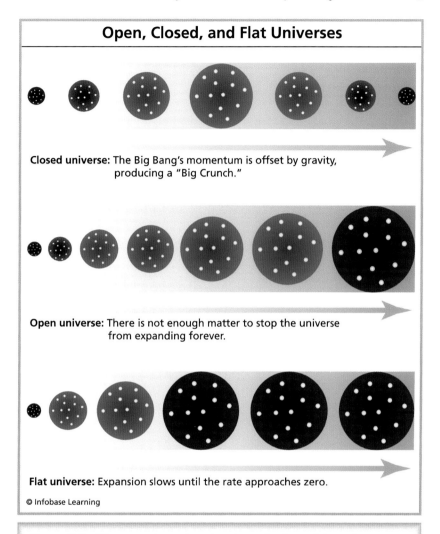

Figure 5.3 There are three theories about the fate of the universe: open, closed, and flat. Scientists' observations have shown that from about 7.5 billion years after the Big Bang onwards, the expansion rate of the universe has been increasing, which reflects the open universe theory.

die, white dwarfs, neutron stars, and black holes would be all that was left over in the universe. The universe would eventually stretch out so far that it would be populated only with dust, gas, and rapidly cooling, decaying stars. Eventually, at about 10^{40} (10 thousand trillion trillion trillion) years from now, scientists predict that even the white dwarfs and neutron stars would disappear when the protons that they are made up of decay. Proponents of this theory believe that, finally, a googol (10^{100}) years or so from now, black holes would dominate the universe.

The last scenario, the big rip, was developed after scientists discovered that the expansion in the universe is accelerating. In the big rip theory, accelerating expansion eventually gets so out of hand that all galaxies, stars, planets, and even atoms and nuclei are torn apart.

Current observations lead scientists to believe that the universe is flat and that the expansion of the universe is speeding up. Depending on the true nature of dark energy, which scientists know little about, this acceleration could eventually slow (and possibly stop) as the model of a flat universe predicts. In this case, scientists believe that the most likely scenario for the end of the universe would be the big chill. On the other hand, dark energy could cause the universe to expand exponentially and, eventually, end in a big rip.

SINGULARITIES AND BLACK HOLES

An expanding universe is not the only unexpected idea predicted by Einstein's general theory of relativity. Shortly after Einstein published his theory, a German astrophysicist named Karl Schwarzschild (1873–1916) used it to determine how a perfectly spherical star's gravity would warp space-time. Just before his death, Schwarzschild was the first to describe a singularity, a region of never-ending curvature in the space-time fabric.

Today, scientists believe that singularities lie at the center of black holes. They also believe that our universe started out as a singularity. Singularities are infinitely dense. They also have infinitely high temperatures. Einstein's equations for general relativity break down when temperature and energy become infinite. Mathematicians use the term singularity to indicate that an equation is no longer working. String theory and M theory are modern theories that strive to improve on Einstein's theory in these circumstances.

Einstein's general theory of relativity can also explain why light cannot escape a black hole. Photons (particles of light) that try to push their way out of the extremely large gravitational field of a black hole lose energy. This loss of energy causes the wavelength of light to shift toward the red end of the electromagnetic spectrum. In 1970, Cygnus X-1 became the first black hole detected in the physical world rather than the theoretical world of mathematics.

Einstein, however, found the idea of both singularities and black holes strange. Even up until his death in 1955, Einstein never did believe either one of these things really existed—even though his own theory predicted that they did.

6

The Big Bang

Scientists now know that the universe is expanding and black holes exist. However, the question remains: How did the universe come to be? In the 1920s, the Russian physicist Alexander Friedmann and Georges Lemaitre (1894–1966), a Belgian cosmologist, proposed a theory to explain how our universe formed. In the 1940s, George Gamow (1904–1968) and his colleages at George Washington University further developed the theory. Today, scientists consider this theory, called the Big Bang theory, the leading concept for describing the birth of the universe. According to the Big Bang theory, at one time there was no universe, space, time, matter, or energy. There was nothing. At some point, however, an infinitely dense, infinitely hot singularity appeared. Scientists believe that our universe started with that singularity about 13.7 billion years ago.

TIMELINE OF THE BIG BANG

Scientists cannot be exactly sure what happened at the very first moment of the Big Bang. The current idea is that the singularity suddenly inflated (the Big Bang itself), expanded, and cooled until it resembled the universe in which we now live.

There is no way for scientists to be exactly certain what happened to cause the sudden inflation of the singularity that eventually became the universe. However, they are pretty sure that the outward expansion from that first instant to about 10^{-43} seconds after the beginning of the universe was exceptionally violent. This is the time period scientists call the Big Bang.

Scientists call the time period between 10^{-43} and 10^{-35} seconds after the Big Bang the Grand Unification Epoch. It is believed that during this epoch, all matter and energy were interchangeable and at equilibrium. In other words, matter was being turned into energy at the same rate energy was being turned into matter. Scientists believe that gravity separated from the other fundamental forces at the beginning of this epoch, but the two nuclear forces and electromagnetism were still one force.

The strong nuclear force was the first of these forces to break away from the others. Scientists believe this occurred at approximately 10^{-35} seconds after the Big Bang. At this time, the universe had cooled a bit, although it was still extremely hot—about 10^{27} Kelvin (water boils at 3.73×10^2 K). The cooling process resulted in a phase change (a little like the phase change that occurs when water freezes and changes into ice), which allowed the strong nuclear force to break away. Scientists call this process symmetry breaking. Symmetry breaking created huge amounts of energy.

THE INFLATIONARY UNIVERSE

In 1980, American physicist Alan Guth (1947–) proposed an idea to clear up some of the inconsistencies in the Big Bang theory. Guth theorized that just a few nanoseconds after the Big Bang (at about

The Big Bang Insult

George Gamow was not the one who named his theory the Big Bang. The term was actually coined by an English astrophysicist named Fred Hoyle (1915–2001) in 1950. Hoyle did not believe in the theory and meant the term to be an insult, but the name stuck anyway.

10^{-35} seconds), the universe inflated. This inflationary period saw the universe expand from a size much smaller than an atom to about the size of a grapefruit in about one-trillionth of a second. To be more specific, the universe expanded a googol (10^{100}) times or more within a billionth of a billionth of a trillionth (10^{-30}) of a second.

Because it happened so fast, this inflation smoothed out the variations in the universe's density. If there was no inflationary period, the universe would not be as uniformly populated with stars and galaxies as it is today. Instead, large clumps would have been left behind. According to this theory, inflation ended 10^{-30} seconds after the Big Bang. Once the period of rapid inflation ended, the energy causing the inflation was converted into dense hot gas. This gas continued to expand and cool, allowing matter to clump together to form stars and galaxies.

Guth based his idea on the Grand Unified Theory (GUT). This theory, based on several different but similar models, predicts that the universe cooled shortly after the Big Bang to allow the forces of gravity, electromagnetism, and the weak and strong forces to separate from each other at phase transitions. If the universe acted the

Misconceptions About the Big Bang

The term *Big Bang theory* seems to indicate that an explosion took place when the universe began, but scientists do not think that is really the case. Instead they believe that, from the singularity, matter and energy suddenly expanded outward at incredible speeds. So to imagine the beginning of the universe, think of an infinitely small balloon being blown up to the size of our universe. The singularity expanded, and time and space expanded inside it. In fact, it is still expanding.

Many people also picture the singularity as a blazing fire-ball that appeared in space. However, before the singularity appeared, there was no space. Space appeared in the singularity, not the other way around. Scientists do not know why the singularity appeared or where it came from.

way that ordinary matter acts at phase changes, Guth argued, there would have been excess energy released that would have counteracted gravity and allowed inflation.

FORMATION OF MATTER

During the time just after the inflationary period, however, the universe was still too hot for atoms to exist. Only plasma, which is a mass of charged atomic nuclei and electrons, could withstand the types of temperatures that existed at that time. Plasma is often called the fourth state of matter. It is similar to a gas, but it exists at much higher temperatures (several thousand to several million degrees Fahrenheit or Celsius) and consists of charged particles instead of neutral atoms. Plasma scatters light so strongly that scientists believe that the early universe was completely opaque.

Eventually, scientists believe that the universe's initial expansion slowed and it cooled even more, allowing first the elementary particles—electrons, quarks, photons, and neutrinos—to condense. From these elementary particles, protons and neutrons started to form and stable atoms of hydrogen and helium could be made. When the plasma started to cool and condense, and the nuclei and electrons were able to join to form neutral atoms, the opaqueness that clouded the universe began to clear. Because there were no free electrons left to scatter the photons of light, the hot, dense fog disappeared and for the first time, the universe became transparent to light.

Over a period of time that lasted roughly a billion years, more and more hydrogen and helium atoms formed. These atoms were attracted to each other and gathered in large clouds of dust and gas. These clouds of dust and gas are called nebulae. About 5 billion years ago, scientists believe that one of these giant interstellar clouds of gas started to compress. Why did it compress? Scientists are not sure exactly, but possibilities include the collision of two nebulae or the possibility that a shockwave from an exploding supernova disturbed the cloud in some way. Whatever caused the collapse of the nebula, however, scientists believe that nuclear reactions started to occur within its hot, dense core and part of the nebula exploded into a blazing star that became our Sun. The rest of the dust and gas in

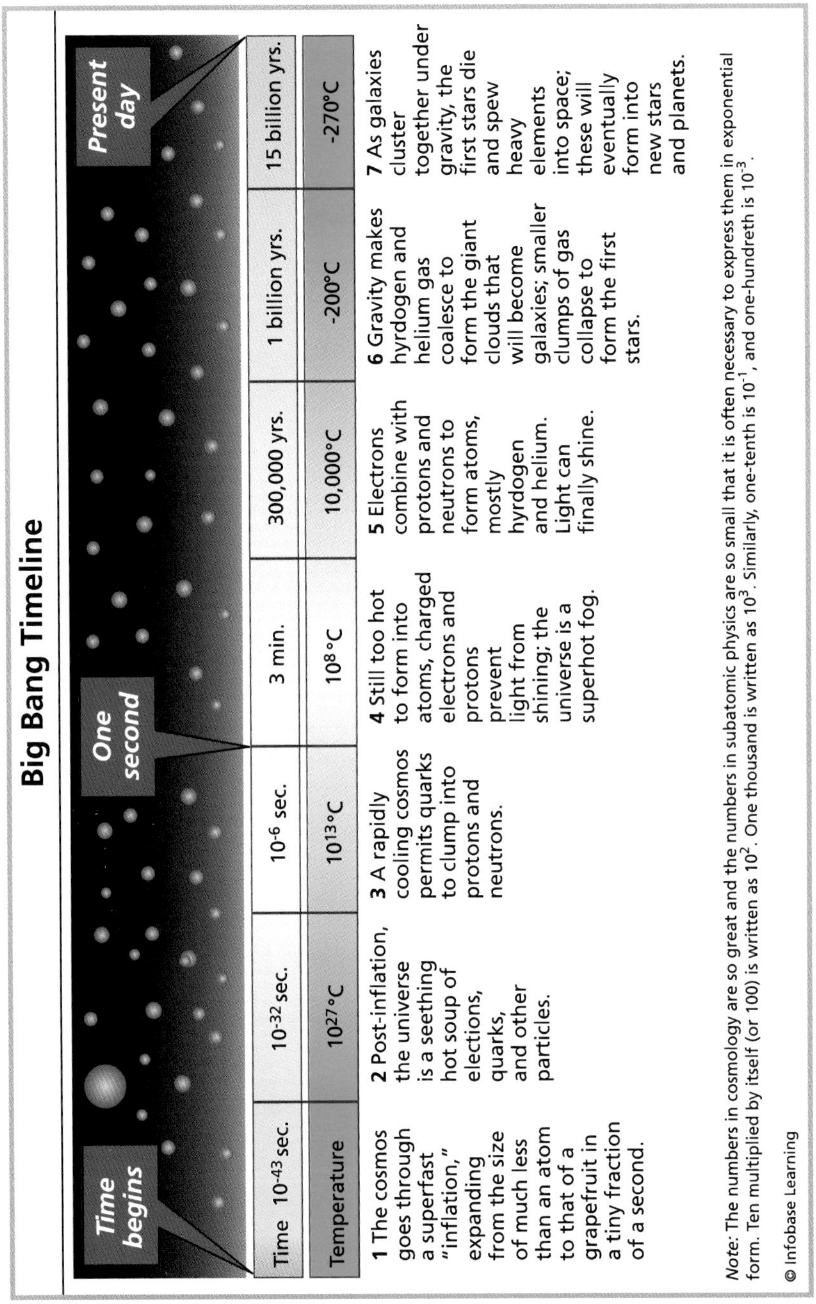

Big Bang Timeline

	Time begins		One second					Present day
Time	10^{-43} sec.	10^{-32} sec.	10^{-6} sec.	3 min.	300,000 yrs.	1 billion yrs.	15 billion yrs.
Temperature		10^{27}°C	10^{13}°C	10^{8}°C	10,000°C	-200°C	-270°C
	1 The cosmos goes through a superfast "inflation," expanding from the size of much less than an atom to that of a grapefruit in a tiny fraction of a second.	2 Post-inflation, the universe is a seething hot soup of electrons, quarks, and other particles.	3 A rapidly cooling cosmos permits quarks to clump into protons and neutrons.	4 Still too hot to form into atoms, charged electrons and protons prevent light from shining; the universe is a superhot fog.	5 Electrons combine with protons and neutrons to form atoms, mostly hyrdogen and helium. Light can finally shine.	6 Gravity makes hyrdogen and helium gas coalesce to form the giant clouds that will become galaxies; smaller clumps of gas collapse to form the first stars.	7 As galaxies cluster together under gravity, the first stars die and spew heavy elements into space; these will eventually form into new stars and planets.

Note: The numbers in cosmology are so great and the numbers in subatomic physics are so small that it is often necessary to express them in exponential form. Ten multiplied by itself (or 100) is written as 10^2. One thousand is written as 10^3. Similarly, one-tenth is 10^{-1}, and one-hundreth is 10^{-3}.

© Infobase Learning

Figure 6.1 According to the Big Bang theory, the universe was originally in an extremely hot, dense state and expanded rapidly. It then cooled by expanding to the present diluted state, and it continues to expand today.

the nebula eventually condensed into the planets and other objects that inhabit our solar system.

EVIDENCE FOR THE BIG BANG

At the moment, scientists believe that the Big Bang model of the universe is the one that fits best with their observations. Several pieces of evidence left behind by the Big Bang exist to support this conclusion. One of them is Hubble's observation that distant galaxies are moving away from us.

Another piece of evidence is radiation left over by the Big Bang. In the beginning, the universe was an incredibly hot place. The best estimates place the temperature at one second after the Big Bang at about 10 billion Kelvin (18 billion degrees Fahrenheit). In 1948, George Gamow, working with his student, Ralph Alpher (1921–2007), predicted that if the Big Bang really did occur, it would have left behind evidence of this intense radiation. Indeed, scientists have found that microwaves, a type of electromagnetic radiation, can still be detected today—13.7 billion years after the universe was created. Astronomers call these microwaves cosmic background radiation (CBR).

This background radiation was discovered by Robert Wilson (1936–) and Arno Penzias (1933–) in 1965. At the time, the two scientists were using the Bell Laboratories radio telescope in Holmdel, New Jersey, for an entirely different reason. They were looking for radio waves that were emitted by objects in space. Wilson was planning to map radio signals emitted by the Milky Way while Penzias was looking for radio waves emanating from the space between galaxies. What they found was what scientists call "noise," a constant background interference similar to the static on a radio or television.

At first, Wilson and Penzias thought that the interference might be caused by pigeons that had been roosting in the antenna, but even after the scientists tried shooing them away and repeatedly scrubbing the radio telescope to remove all the droppings, the noise persisted. Trying to find another possible source of the noise, they pointed the antenna directly toward nearby New York City. However, there was absolutely no change in the amount of static detected, so they ruled out interference from that direction. Wilson and Penzias continued to search for the origin of the noise throughout the year and found that no matter what the season, it stayed the same. After ruling out all the ordinary reasons for the interference that they could think of (including those coming from Earth and the solar system), they started to look for other sources. At this point, they got in touch with Robert Dicke (1916–1997), a physicist at nearby Princeton University.

Dicke very quickly realized that Wilson and Penzias had found evidence of the Big Bang. The noise was the remains of the heat created during the infant days of the universe. The heat was detected as microwaves because the universe is expanding. The wavelength of the radiation would have redshifted and weakened between the time of the Big Bang and when it was detected in 1965. Today, this cosmic background radiation (CBR) has a temperature of about -270°C (2.7 K) and has a wavelength of a few millimeters. This wavelength is in the microwave section of the electromagnetic spectrum. When the radiation was first emitted, however, its wavelength was a thousand times shorter—only a few microns—and it was emitted as infrared radiation, or heat. The microwave radiation discovered by Wilson and Penzias indicates that the universe was once a much hotter and denser place than it is today. Wilson and Penzias were awarded the Nobel Prize in physics in 1978 for their discovery.

THE STEADY-STATE THEORY

Dicke may have been convinced that Wilson and Penzias had discovered evidence of the Big Bang, but Wilson was not. He was trained in a conflicting theory that became popular in the 1950s, which was called the steady-state theory. The steady-state theory of the universe stated that the universe had always existed and looked the way it does today and that it would forever remain that way. This theory was developed by Hermann Bondi (1919–2005), Thomas Gold (1920–2004), and Fred Hoyle in 1948, the same year that Gamow developed the Big Bang theory.

If the universe had always existed in a steady state, however, Hubble's observations would have to be explained. Steady-state proponents said that Hubble's observations could fit within their theory if matter was continuously being created to take the place of the

Figure 6.2 Robert W. Wilson and Arno A. Penzias, Bell Lab employees who won the 1978 Nobel Prize in physics, are shown standing in front of their microwave antenna at Bell Labs in Holmdel, New Jersey in 1978. They shared the prize for their discovery of long-wavelength, low-temperature radio emissions from the universe.

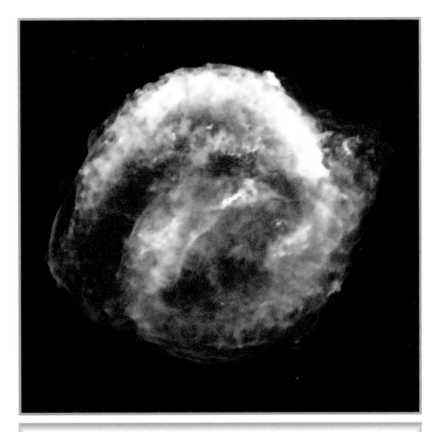

Figure 6.3 Scientists created an image of one of the youngest supernova remnants in the galaxy using NASA's Chandra X-ray Observatory. In this image, red represents low-energy X-rays and shows material around the star—dominated by oxygen—that has been heated up by a blast wave from the star's explosion. Yellow denotes slightly higher energy X-rays, mostly iron formed in the supernova, while green (medium-energy X-rays) shows other elements from the exploded star. Blue represents the highest energy X-rays and shows a shock front generated by the explosion.

matter that was spreading outward. If this was what was truly going on, by the time the distance between galaxies doubled, there would be enough matter in between the galaxies to form another. This idea is not as far-fetched as it may sound. The amount of matter formed would be tiny—only about one or two atoms every year for each cubic mile.

Another question proponents of the steady-state theory had to answer was where did the heavier elements that are found all around us come from. They answered it by looking at the stars. At the center of most stars, temperatures are in the millions of kelvin. Stars make energy by fusing hydrogen into helium. The energy eventually makes its way to the surface of the star and is given off as light. This outward radiation of energy counterbalances the inward force of gravity.

However, eventually, stars run out of hydrogen fuel. When this happens, the force of gravity starts to crush the star's center which, in turn, increases the core temperature (to about 100 million kelvin). At these hotter temperatures, new fusion reactions can take place and heavier elements such as carbon, oxygen, and other elements that are needed to sustain life can be made. The reactions release a lot of energy in the process, too. Massive stars have enough mass and energy for this crushing of the core to take place at even higher temperatures. At these temperatures, even heavier elements like neon, silicon, and even iron, can be formed. Forming iron, however, takes a lot of energy—often more energy than a star can produce—allowing gravity to take over and the star to collapse.

Is Anyone Out There?

In February 2008, astronomers from the University of Arizona announced at the meeting of the American Association for the Advancement of Science that their research shows that it is likely that there is life out there somewhere in the universe. According to their calculations, between 20% and 60% of the yellow dwarf stars in the universe are likely to have rocky, terrestrial planets like Earth orbiting them. In the future, scientists hope to be able to detect planets that reside in what they call "the Goldilocks zone," which is defined as an area far enough from its Sun so that it is not too hot, but close enough so that it is not too cold for life to thrive. Those terrestrial planets may have life on them just as Earth does.

Eventually, the star goes supernova. The explosion of the supernova scatters heavy elements throughout space where they can clump together and become additional stars, planets, or possibly even life forms. In fact, Earth is constantly being bombarded with the leftovers of supernova explosions. These high-energy particles, called cosmic rays, travel at speeds equal to approximately 90 to 99% of the speed of light.

The debate between steady-state supporters and Big Bang theory proponents went on throughout the 1940s and 1950s. However, supporters of the steady-state theory were unable to explain several observations. One of the things that they could not account for was the amount of helium found in the universe. If the steady state theory were true, there would be a lot less helium in the universe today than data indicates. The theory also could not account for the cosmic background radiation found by Wilson and Penzias. The Big Bang theory, however, can account for both of these observations.

Not long after Wilson and Penzias found the CBR, Stephen Hawking, along with Roger Penrose (1931–), was able to prove that, mathematically, Einstein's general theory of relativity showed that the universe should have started with a singularity and that it will likely end in a black hole. If the universe did, indeed, begin as a singularity, a Big Bang must have followed. Hawking's calculations, the cosmic background radiation, and evidence that distant galaxies contain many more quasars than closer galaxies convinced many scientists that the universe did not exist in a steady state. Instead, it had evolved over time and today's universe looks very different from the beginning universe. By the 1970s, most scientists decided that the Big Bang theory was the best model of the universe that had been developed so far.

The theory did correctly predict three things that scientists have now been able to measure or observe. First, the Big Bang theory predicted that the universe should be expanding. Edwin Hubble proved that it was. Second, the theory predicted that about 25% of mass in the universe should be the helium created during the first few minutes after the Big Bang, and scientists have observed this to be true, too. And, lastly, the theory predicted the presence of cosmic background radiation, which Wilson and Penzias proved to exist.

The Future Universe

Not everything is understood about the beginning of the universe and what may or may not have happened at that time. As time goes on, scientists may need to modify the Big Bang theory, or come up with a new theory, to account for new observations.

ADDITIONS TO THE BIG BANG THEORY

Because scientists cannot recreate the extreme conditions that existed at the beginning of the universe in the laboratory, theory is the only thing that can guide our thinking about that time. The Big Bang theory does a good job of explaining a lot about the universe, but it breaks down when it comes to explaining the evolution of the earliest part. Any theory that tries to explain how the Big Bang came about will have to incorporate the observations that have been made so far. Two problems that scientists have encountered with the Big Bang theory—the horizon problem and the flatness problem—have already been addressed.

The horizon problem shows itself in the uniformity of the cosmic background radiation. Wherever in the universe that cosmic background radiation has been detected, it varies by less than one part per ten thousand. The temperature of this radiation is now a uniform 2.73 K (-454.8°F). In order for the temperature to be the same, particles in the universe must collide with each other on a regular basis.

However, the east and west edges, or the horizon, of our universe today are separated by about 10 million light-years. How can such far-flung regions have the same temperature? Alan Guth proposed the inflationary theory to address this problem.

The inflationary theory also helped scientists explain the observations that our universe is, essentially, flat. Scientists have observed that matter is relatively uniform throughout the known universe. Before the inflation theory, observations that the universe was flat did not fit with the standard Big Bang theory. Without inflation, there would be clumps of matter rather than more uniformly spread matter across the universe as it is seen today. This problem was called the flatness problem.

Scientists eventually accepted the addition of the inflation theory to the standard Big Bang model to address these problems. The inflationary theory explains how the matter and energy in the universe were once in extremely close contact, and, therefore, at the same temperature. Through rapid inflation, this temperature remained constant all across the universe and clumps of matter were smoothed out. Still, the inflationary theory has its problems as well. For example, what stopped the inflation? So far, the answer to that question remains unclear.

Other additions to the Big Bang theory help explain dark matter and dark energy. In the 1940s, when scientists first developed the Big Bang theory, they only knew about ordinary matter. They knew nothing of dark matter or dark energy, so the theory did not include either of these elements. Dark matter was introduced into the theory in the 1980s. Later, in the 1990s, dark energy and the idea that the universe's expansion is accelerating were also worked into the theory. However, scientists still do not know exactly why the expansion seems to be accelerating.

OTHER THEORIES

The Big Bang theory is not the only theory of the universe. Paul Steinhardt (1952–), the Albert Einstein Professor of Science at Princeton University and a theoretical physicist, and Neil Turok (1958–), the chair of mathematical physics at Cambridge University in England, for example, have a different theory.

The Future Universe 79

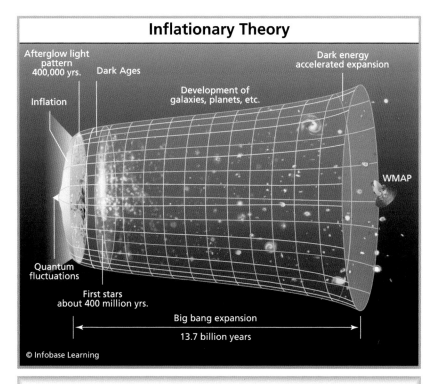

Figure 7.1 The Inflationary Theory proposes that there was a period of extremely rapid expansion of the universe during its first few moments.

Steinhardt and Turok agree that there are all kinds of evidence pointing to our universe being created from some hot, dense core some 14 billion years ago. They also agree that it has been expanding and cooling ever since. They differ with other scientists, however, in that they do not believe that physics will someday find a way to explain how and why the Big Bang occurred, much less how the universe is going to end.

Steinhardt and Turok believe that it is possible that the Big Bang was not the beginning of space and time at all. Instead, it is possible that our universe goes though multiple cycles of evolution. They believe that each cycle influences the next cycle. This cyclic, or oscillating, model of the universe proposes that the Big Bang was not a one-time event. Instead, in each cycle, a Big Bang creates hot matter and radiation. Expansion and cooling allow the stars and galaxies to form and, after about a trillion years, the universe collapses into a big crunch. After the big crunch, another Big Bang takes place and starts the entire cycle over again.

Another theory called the no-boundary universe has been proposed by Stephen Hawking and Jim Hartle. In this theory, the universe is finite in size and did not start out as a singularity. The geometry of a no-boundary universe is much like that of a globe. If, for example, you start at the North Pole and trace a finger all the way around the globe and back to the North Pole, your finger would never encounter an edge. A no-boundary universe is similar. There is a starting point (the North Pole), but there is no singularity (just as the North Pole is not a singularity). The mathematics of this theory is very complicated and it involves what physicists call imaginary time (similar to imaginary numbers in mathematics). Only time will tell if Steinhardt and Turok's endless universe or Hawking and Hartle's no-boundary universe have any merit. For now, most scientists still believe the standard, conventional Big Bang theory is the best scientific model of the universe that they have.

INSTRUMENTS USED BY ASTRONOMERS

To learn all they can about the current state of the universe, scientists use powerful telescopes and technology. The technologies that scientists have at their fingertips today have allowed them to observe many of the phenomena that resulted in the theories they have developed. To learn even more about the universe, however, research into even more powerful equipment will be necessary.

Telescopes and Space Probes

The Hubble Space Telescope was sent into space in 1990. Because the telescope is outside of Earth's atmosphere, the light that comes from distant stars and galaxies is less distorted. The telescope's light-gathering abilities allow astronomers to see celestial objects much farther away then they ever have before.

Between December 18 and 28, 1995, the Hubble Space Telescope was used to take a total of 342 pictures of a spot of sky within the

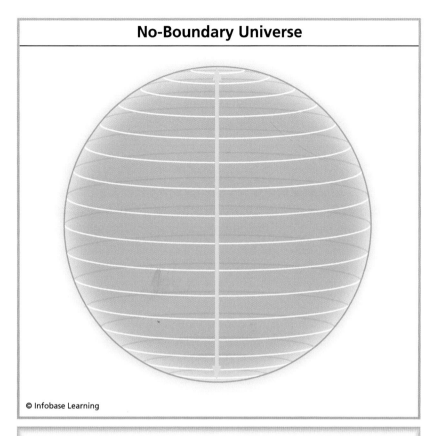

Figure 7.2 The no-boundary universe proposal is Stephen Hawking's attempt to combine the ideas of general relativity and quantum mechanics. In this theory, space and time have neither a beginning nor an end and, therefore, the universe did not start out as a singularity.

constellation Ursa Major. The resulting composite of photographs is called the Hubble Deep Field (HDF). The HDF photographs not only captured the visible light part of the electromagnetic spectrum, but also the ultraviolet and infrared spectrums. Telescopes before Hubble had never revealed another galaxy in this area, so scientists were not really expecting to see much—at the most, a faint, fuzzy picture of a few galaxies. Instead, they found an astonishing 2,000 galaxies.

The brightest objects in the Hubble Deep Field images are about 7 to 8 billion light-years from Earth. Therefore, scientists are actually seeing those galaxies as they were 7 to 8 billion years ago. Some of the more faint shadows in the Hubble Deep Field can show astronomers what the universe was like around 12 billion years ago.

In March 2004, another study of the farthest reaches of the universe was unveiled. This set of photographs taken by the Hubble Telescope is called the Hubble Ultra Deep Field (HUDF). By studying the HUDF, astronomers now know that the universe is made up of an incredible number of galaxies. For example, at one time, scientists thought that the entire cup of the Big Dipper might contain upwards of 400 galaxies. However, the HUDF photographs revealed that an area of space just three-thousandths the size of the cup of the Big Dipper has at least 10,000 galaxies in it. That means that the cup of the Big Dipper really contains well over 190 million galaxies. Furthermore, as telescopes become more powerful, chances are that many more galaxies will be found one day.

In 1989, about six months before the Hubble Telescope was sent into orbit, the U.S. National Aeronautics and Space Administration (NASA) launched the Cosmic Background Explorer (COBE). Designed to collect data about the nature of cosmic background radiation, COBE provided some of the information scientists needed to better understand the early universe. The leaders on that mission, John Mather (1946–) of the Goddard Space Flight Center and George Smoot (1945–) of the University of California at Berkeley, were awarded the 2006 Nobel Prize in Physics for their work on this mission. In 1992, COBE was again used in the first attempt to map the slight variations in the cosmic background radiation temperature.

In 2001, NASA launched a satellite designed to take high-resolution, whole-sky images of the dark layer of space at the outermost edge of today's visible universe. The satellite is called the Wilkinson Microwave Anisotropy Probe (WMAP). The images from the WMAP contained the evidence that finally allowed scientists to determine that, on a large scale, the universe is essentially flat.

Particle Accelerators

Telescopes are not the only instruments scientists use to help them study the universe. Particle accelerators, which are sometimes called "atom smashers," are extremely helpful in studying elementary particles, looking for clues to understanding dark matter and

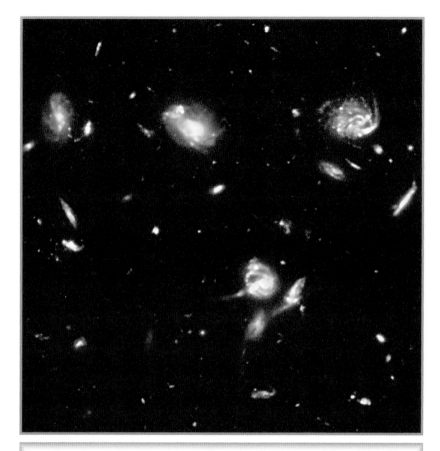

Figure 7.3   Galaxies are shown colliding in this Hubble Ultra Deep Field image.

dark energy, recreating conditions that occurred during the Big Bang, as well as conducting many other types of experiments. The few large particle accelerators that are now in existence around the world are shared by teams of international scientists for experiments.

Inside a particle accelerator, particles, such as protons, can be made to travel very close to the speed of light. Scientists make some of the particles travel clockwise around the circle of the accelerator while other particles are forced to travel counterclockwise. When two particles collide, which they often do, some of the energy released turns to matter. The matter created can sometimes be an elementary particle that has never been seen before. It is possible that some of these elementary particles are the same type of particles that were present when the Big Bang occurred. However, particles made in a particle accelerator do not last for long. They appear and then disappear in a matter of nanoseconds.

Scientists at Brookhaven National Laboratory in New York have used their particle accelerator, which is called the Relativistic Heavy Ion Collider, to create the quark-gluon plasma that scientists believe formed just after the Big Bang. Using Germany's Hadron Electron Ring Accelerator (HERA), scientists discovered the particle known as a pentaquark. This particle contains five quarks and the hard-to-find "charm" quark. Scientists were able to make about 50 of the new elementary particles. These are just two examples of what scientists have found so far during the millions of particle collisions they have created over many years.

Scientists continue to use their old particle accelerators, but they are building new ones, as well. The world's largest and highest-energy particle accelerator, the Large Hadron Collider, located on the border of Switzerland and France, was tested for the first time in September 2008. It contains more than 1,600 magnets that weigh more than 30 tons (27.2 metric tons) apiece and are arranged in a 17-mile (27.3-km) circle. The magnets are used to keep the particles moving around the circle. Without them, the particles would just continue in a straight line—at a speed of 99.9999991% of the speed of light. Scientists are hoping that accelerating the particles that close to the speed of light will create collisions that are as energetic as the ones that occurred during the Big Bang. This could yield new elementary particles that have never been seen before and provide the opportunity to study them.

Supercomputers

Solving Einstein's equations, simulating the Big Bang, and sifting through huge amounts of images and data generated by telescopes, particle accelerators, and other equipment takes a lot of computing power. Research centers around the world have developed powerful software programs to process the data and model the cosmos. To run the software, their computers must be able to handle billions, or even trillions, of calculations per second.

Figure 7.4 In this image of the Large Hadron Collider, eight toroid magnets surround the calorimeter. When moved to the middle of the detector, the calorimeter measures the energies of particles produced when protons collide.

THE FUTURE OF COSMOLOGY

There are many questions that scientists hope to be able to answer in the future. Using the Large Hadron Collider, for example, physicists might finally be able to answer the question of what exactly dark matter is. Currently, scientists believe that the fundamental parts that they are so familiar with today—the electron, muon, and quark, for example—had much heavier counterparts in the early universe. Physicists call the heavier counterpart of an electron the selectron. A muon's counterpart would be a smuon, and a quark's counterpart would be a squark. This is a theory called supersymmetry. The question is, though, can scientists create these particles? Furthermore, can they retain them long enough to study and answer some questions about them? Could any of these particles have been stable enough to survive until the present time? This is possible, but if they have survived, scientists have been unable to detect them so far.

Another particle that scientists hope to create in a particle accelerator some day is the Higgs boson, named for the English physicist Peter Higgs (1929–) who suggested its existence in the 1960s. Higgs, and most other physicists, believe that this particle carries a field that interacts with all other particles. Scientists believe that the Higgs field is what gives other elementary particles, like electrons and quarks, mass. If these particles did (or do) exist, scientists expect that they are 100 to 200 times more massive than a proton. At least, that is the theory. It will not be an easy task to produce and gather evidence about these particles. Large particles like the Higgs tend to decompose into smaller particles very quickly—fractions of seconds, in fact. Large particles also require large amounts of energy to produce them, which, in turn, means that a very large, high-energy supercollider like the Large Hadron Collider must be used. Scientists do not expect that the first or even the millionth collision on the Large Hadron Collider will result in the production of a Higgs particle. In fact, they think it may take trillions of collisions to produce just one of them. And chances are they will not even realize that they have been successful when it happens (if it ever does). Only the collider's detectors will know. The scientists will not become aware of their success until they wade through a huge amount of the detector's computer data. In addition, the detector would not be able to actually detect the Higgs particle itself. Instead, it can only identify the products from the breakdown of the particle. In other words, if a Higgs particle is ever produced, only the fact that it was once present and that it is already in the process of disappearing will be known.

Why do scientists care about any of this? Well, theoretically, the Big Bang should have produced an equal amount of matter and antimatter. If that did happen, these particles would have annihilated each other every time they collided, and the universe would be, well, empty. However, the universe is not empty. In addition, there is evidence of the existence of some mass and energy that we cannot see. What types of mass and energy produced this observational data? Where did they come from? These are all questions that scientists hope to answer some day.

Technology has allowed us to see deeper into the universe than ever before. Innovations in telescopes, light-detecting technology, and computing power to model more and more complex information will allow astronomers to see more and more of what exists in the

The Theory of Everything

The general theory of relativity explains much about our universe. It shows how light bends as it passes a massive object and describes how planets remain in orbit around a sun and how our universe works on a large scale. It even predicted that our universe was expanding even when most scientists, including Einstein himself, believed that this was impossible. Yet general relativity cannot explain what the singularity at the beginning of our universe was like. At that point, Einstein's equation breaks down. To look back further than 10^{-43} seconds after the universe was born, scientists need a different theory—one that will combine general relativity and quantum mechanics (the theory that explains subatomic interactions).

Scientists call this unified theory the "Theory of Everything." They hope that one day, the way that the weak force, strong force, electromagnetic force, and gravity interact will be described mathematically. So far, they have developed the Grand Unified Theory (GUT) that shows how the weak force, the strong force, and the electromagnetic force are linked mathematically. They have not yet figured out what role gravity plays in these interactions, however. Einstein spent the later years of his life looking for a unified theory, but never found it. In fact, scientists are still searching for the "Theory of Everything" today.

String theory is one of the theories that scientists are looking into to help unite Einstein's general theory of relativity and quantum mechanics. String theory proposes that everything is made up of vibrating strings instead of particles. Each string is tiny, only about 10^{20th} of a proton (100 billion billion times smaller than a proton). And each string vibrates in a 10-dimensional space-time. Scientists believe that the differences between the different particles they see are caused by the different configurations of these strings.

Only time and more research will determine if string theory is actually the theory of everything. Before that can happen, the theory's predictions must be verified through research and observation. At the moment, however, string theory is the best one scientists have.

universe and help them to determine how galaxies like ours grow and evolve over time.

To determine the ultimate fate of the universe, scientists need to know more about dark matter, dark energy, and gravity. The amount of dark matter and the strength of dark energy in comparison to gravity is critical to whether or not the universe will continue to expand and how fast. For now, astronomers believe the universe will continue its accelerating expansion. Still, who knows what new piece of information may be found in the next century, the next decade, or even next year that could change that view.

Glossary

atom   Tiny particles that all matter is made of

Cepheid   A type of star whose brightness varies regularly over time

dark energy   Energy that cannot be detected, but that seems to be speeding up the expansion of the universe

dark matter   Matter that does not give off enough radiation for scientists to detect it

electromagnetic force   The fundamental force that holds atoms together

electron   Negatively charged subatomic particles found traveling around the nucleus of an atom in energy levels

element   Chemical substances that cannot be broken down by ordinary chemical means

fundamental particles   Subatomic articles that cannot be broken down into smaller particles; also called elementary particles

galaxy   A group of stars, dust, and gas that is held together by gravity

gravitational lensing   A phenomenon that occurs when an object is so massive that it bends light

gravity   A weak fundamental force that attracts one massive body to another

isotope   Atoms of the same element that have a different number of neutrons

light-year   The distance that light travels in one year

matter Anything that has mass and takes up space

nebula An interstellar cloud of dust and gas

neutron Neutral subatomic particles found in the nucleus of an atom

nuclear fusion A process in which the nuclei of small atoms fuse to make the nuclei of larger atoms

nucleus The central core of an atom that is made up of protons and neutrons

proton Positively charged subatomic particles found in the nucleus of an atom

quarks Fundamental particles that make up protons and neutrons

quasar Objects at the center of distant galaxies that send out powerful radio signals

retrograde motion The observations that certain celestial objects seem to travel backward

strong force The fundamental force that holds the nucleus of atoms together

subatomic particles Particles that are smaller than an atom

supernova The explosion of a massive star when it dies

wavelength The distance from the top of one wave to the top of the next

weak force The fundamental force responsible for the nuclear decay of some elements and subatomic particles

Bibliography

Achenbach, Joel. "The God Particle." Mar. 2008. National Geographic Magazine. Available online. Accessed Dec. 21, 2008. URL: http://ngm.nationalgeographic.com/2008/03/god-particle/achenbach-text/1.

Alcatel-Lucent. "Cosmology–Penzias and Wilson's Discovery is One of the Century's Key Advances." Available online. Accessed Dec. 21, 2008. URL: http://www.bell-labs.com/project/feature/archives/cosmology/.

AllAboutScience.org. "Big Bang Theory." Available online. Accessed Dec. 21, 2008. URL: http://www.big-bang-theory.com/.

ALS MND Alliance. "What Is ALS/MND?" Available online. Accessed Dec. 21, 2008. URL: http://www.alsmndalliance.org/whatis.html.

Arnett, Bill. "The Nine Planets Solar System Tour." Available online. Accessed Dec. 21, 2008. URL: http://www.nineplanets.org/nineplanets.html#toc.

Berman, Bob. *Strange Universe*. New York: Henry Holt and Co., 2003.

British National Space Centre. "BNSC Learning Zone." Available online. Accessed Dec. 21, 2008. URL: http://www.bnsc.gov.uk/3261.aspx.

Carroll, Rebecca. "First Pictures of Alien Planet System Revealed." Nov. 13, 2008. National Geographic News. Available online. Accessed Dec. 21, 2008. URL: http://news.nationalgeographic.com/news/2008/11/081113-planet-pictures.html?source=rss.

East Tennessee State University. "Black Holes." Available online. Accessed Dec. 21, 2008. URL: http://www.etsu.edu/physics/bsmith/blackholes/bh.html.

European Space Agency. "Introduction to the Universe." Available online. Accessed Dec. 21, 2008. URL: http://sci.esa.int/science-e/www/object/index.cfm?fobjectid=35696&fbodylongid=1682.

The Franklin Institute. "The Case Files." Available online. Accessed Dec. 21, 2008. URL: http://www.fi.edu/learn/case-files/index.php.

Holladay, April. "Models of the Solar System and in Defense of the Fly." USATODAY.com. Available online. Accessed Dec. 21, 2008. URL:

http://www.usatoday.com/tech/columnist/aprilholladay/2005-08-18-solar-model-flies_x.htm.

Howstuffworks Science Channel. "How are Astronomers Able to Measure How Far Away a Star Is?" Available online. Accessed Dec. 21, 2008. URL: http://science.howstuffworks.com/question224.htm.

Internet Encyclopedia of Science. "Astronomers and Astrophysicists." Available online. Accessed Dec. 21, 2008. URL: http://www.daviddarling.info/encyclopedia/A/astronomers_and_astrophysicists.html.

Jones, Terry Jay and Jeanne Hanson. *Astronomy for the Utterly Confused*. New York: McGraw-Hill, 2007.

Kanipe, Jeff. *Chasing Hubble's Shadows: The Search for Galaxies at the Edge of Time*. New York: Hill and Wang, 2006.

NASA. "WMAP Introduction to Cosmology." Available online. Accessed Dec. 21, 2008. URL: http://map.gsfc.nasa.gov/universe/.

NASA's Amazing Space. "Telescopes from the Ground Up." Available online. Accessed Dec. 21, 2008. URL: http://amazing-space.stsci.edu/resources/explorations//groundup/.

NASA's Goddard Space Flight Center. "Imagine the Universe Science." Available online. Accessed Dec. 21, 2008. URL: http://imagine.gsfc.nasa.gov/docs/science/science.html.

The Night Sky This Week. "The Hyades Star Cluster." Available online. Accessed Dec. 21, 2008. URL: http://www.nightskyinfo.com/archive/hyades/.

PBS. "A Science Odyssey." Available online. Accessed Dec. 21, 2008. URL: http://www.pbs.org/wgbh/aso/.

PBS. "Stephen Hawking's Universe." Available online. Accessed Dec. 21, 2008. URL: http://www.pbs.org/wnet/hawking/html/home.html.

PBS/NOVA. "Einstein's Big Idea." Available online. Accessed Dec. 21, 2008. URL: http://www.pbs.org/wgbh/nova/einstein/.

Rice University. "The Galileo Project." Available online. Accessed Dec. 21, 2008. URL: http://galileo.rice.edu/index.html.

Sloan Digital Sky Survey. "Basic Science Projects." Available online. Accessed Dec. 21, 2008. URL: http://cas.sdss.org/dr6/en/proj/basic/.

Smith, Scott. "GEO 154 Homepage." La Salle University. Available online. Accessed Dec. 21, 2008. URL: http://www.lasalle.edu/~smithsc/Astronomy/astro.html.

Southwest Research Institute. "Plasma: the 4th State of Matter." Available online. Accessed Dec. 21, 2008. URL: http://pluto.space.swri.edu/image/glossary/plasma.html.

SPACE.com. "The Big Rip: New Theory Ends Universe by Shredding Everything." Available online Accessed Dec. 21, 2008. URL: http://www.space.com/scienceastronomy/big_rip_030306.html.

Steinhardt, Paul and Neil Turok. *Endless Universe: Beyond the Big Bang.* New York: Doubleday, 2007.

University Corporation for Atmospheric Research. "Windows to the Universe." Available online. Accessed Dec. 21, 2008. URL: http://www.windows.ucar.edu/.

University of Colorado at Boulder. "The George Gamow Memorial Lecture Series." Available online. Accessed Dec. 21, 2008. URL: http://www.colorado.edu/physics/Web/Gamow/career.html.

University of Illinois' National Center for Supercomputing Applications. "Cosmic Mystery Tour." Available online. Accessed Dec. 21, 2008. URL: http://archive.ncsa.uiuc.edu/Cyberia/Cosmos/CosmicMysteryTour.html.

University of Illinois' National Center for Supercomputing Applications. "Cosmos in a Computer." Available online. Accessed Dec. 21, 2008. URL: http://archive.ncsa.uiuc.edu/Cyberia/Expo/cosmos_nav.html.

University of Illinois' National Center for Supercomputing Applications. "Spacetime Wrinkles." Available online. Accessed Dec. 21, 2008. URL: http://archive.ncsa.uiuc.edu/Cyberia/NumRel/NumRelHome.html.

University of Oregon's Department of Physics. "Astronomy Hypertext Book." Available online. Accessed Dec. 21, 2008. URL: http://zebu.uoregon.edu/text.html.

University of Tennessee Department of Physics & Astronomy. "Astronomy 161: The Solar System." Available online. Accessed Dec. 21, 2008. URL: http://csep10.phys.utk.edu/astr161/lect/index.html.

University of Texas McDonald Observatory. "Black Hole Encyclopedia." Available online. Accessed Dec. 21, 2008. URL: http://blackholes.stardate.org/.

Further Resources

Berger, Melvin. *Did It Take Creativity to Find Relativity, Albert Einstein?* New York: Scholastic, Inc., 2007.

Graham, Ian. *Earth and the Universe.* New York: Children's Press, 2008.

Hawking, Stephen and Lucy Hawking. *George's Secret Key to the Universe.* Waterville, Maine: Thorndike Press, 2008.

Jackson, Ellen. *The Mysterious Universe: Supernovae, Dark Energy, and Black Holes.* Boston, Mass: Houghton Mifflin Harcourt, 2008.

Kerrod, Robin and Jayne Parsons. *The Way the Universe Works.* New York: DK Publishing, Inc., 2002.

Kirkland, Kyle. *Particles and the Universe.* New York: Facts on File, Inc., 2007.

Stone, Lynn. *Space.* Vero Beach, Fla: Rourke Publishing, LLC, 2008.

Venezia, Mike. *Albert Einstein: Universal Genius.* Danbury, Conn.: Scholastic Library Publishing, 2009.

———. *Stephen Hawking: Cosmologist Who Gets a Big Bang Out of the Universe.* Danbury, Conn: Scholastic Library Publishing, 2009.

Yount, Lisa. *Modern Astronomy: Expanding the Universe.* New York: Chelsea House Publishers, 2006.

Web Sites

Enchanted Learning Astronomy
http://www.enchantedlearning.com/subjects/astronomy/
> Enchanted Learning's Astronomy page provides some quick facts about the Milky Way.

PBS Astronomy
http://www.pbs.org/seeinginthedark/astronomy-topics/
> For a more in-depth look at the planets in our solar system as well as those in others, try PBS's Astronomy page.

Further Resources

NOVA's "Einstein's Big Idea"
http://www.pbs.org/wgbh/nova/einstein/
> "Einstein's Big Idea" was a NOVA television program aired on PBS stations in 2005. This Web site provides an overview of the show as well as related articles and interactive resources to help students understand Einstein's ideas.

NOVA's "Elegant Universe"
http://www.pbs.org/wgbh/nova/elegant/
> Have questions about string theory and its 10 dimensions? NOVA's "Elegant Universe" has the answers. This companion Web site allows you to watch the 3-hour television program that explains this theory, read articles about physicists' search for the theory of everything, and explore images from atom smashers.

NOVA's "Hunting the Edge of Space"
http://www.pbs.org/wgbh/nova/space/hunting-edge-space-1.html
> This NOVA television program, originally aired April 6, 2010, explores the history of telescopes and how they continue to change the way scientists view our universe. The first hour is available online.

NOVA's "Biggest Machine Ever Built"
http://www.pbs.org/wgbh/nova/physics/large-hadron-collider.html
> Explore articles, audio, and expert Q & A about the Large Hadron Collider.

NOVA's "Dark Matter"
http://www.pbs.org/wgbh/nova/physics/dark-matter.html
> Watch NOVA's hour-long program that explains what scientists have discovered about dark matter so far.

NASA Science
http://science.nasa.gov/
> Interested in the science behind the Big Bang, dark energy, and dark matter? NASA Science page explains it all.

NASA: Grades 9–12
http://www.nasa.gov/audience/forstudents/9-12/index.html
> Looking for more up-to-date infromation on NASA missions, career profiles, or ways you can become involved with the space program? Check out NASA's site specifically designed for teens.

Picture Credits

Page

10: © Infobase Learning
11: © Infobase Learning
14: © David Parker/Photo Researchers, Inc.
17: © Courtesy of the Archives, California Institute of Technology
18: © Infobase Learning
23: © Infobase Learning
25: © NASA Marshall Space Flight Center Collection
26: © Infobase Learning
29: © Infobase Learning
30: © Gemini Observatory/ Association of Universities for Research in Astronomy
32: © NASA Jet Propulsion Laboratory Collection
35: © Infobase Learning
36: © Infobase Learning
38: © World History Archive/ Alamy
41: © Mary Evans Picture Library/Alamy
43: © AP Images/NASA, Paul E. Alers
44: © SSPL via Getty Images
46: © Infobase Learning
49: Library of Congress Prints and Photographs Division
50: © Infobase Learning
54: © Infobase Learning
59: © Infobase Learning
60: © Infobase Learning
63: © Infobase Learning
70: © Infobase Learning
73: AP Images
74: NASA/CXC/NCSU/ S.Reynolds et al
79: © Infobase Learning
81: © Infobase Learning
83: NASA, ESA, S. Beckwith (STScI) and the HUDF Team
85: © Maximilien Brice/CERN

Index

A

accelerating universe theory, 61–64
accelerators, particle, 81–83, 85
Alpher, Ralph, 71
amyotrophic lateral sclerosis (ALS), 42–43
Andromeda Galaxy, 22
antimatter
- dark matter vs., 19
- overview of, 14
- special theory of relativity and, 50

antiquarks, 14
Aristarchus of Samos, 36
armless spiral galaxies, 24
asteroids, 31
astronomical units (AU), 21–22
astronomy, history of
- in ancient times, 32–33
- Brahe and, 36–38
- Copernicus and, 34–36
- Einstein and, 47–52
- Galileo and, 39–41
- Hawking and, 42–43
- Kepler and, 38–39
- Newton and, 42–46
- overview of, 32
- Ptolemy and, 33, 34
- Shapley and, 46–47

atom smashers, 82–84
atomic mass units (amu), 8
atoms, 7–9

B

Babylonians, 33
barred spiral galaxies, 24
Big Bang theory
- additions to, 77–78
- evidence for, 71–72
- formation of matter and, 69–71
- inflationary universe and, 67–69
- misconceptions about, 68
- overview of, 66
- steady-state theory vs., 72–75
- timeline of, 66–70

big chill scenario, 62–64
big crunch scenario, 62
big rip scenario, 62, 64
binary stars, 28
black holes, 27–28, 29, 53, 64–65
blueshift, 58, 59
Bondi, Hermann, 72
bottom quarks, 9, 11
Brahe, Tycho, 37–39
A Brief History of Time (Hawking), 42

C

Cassiopeia, 37
Catholic Church, 36, 40–41
CBR. *See* Cosmic background radiation
Cepheid variable stars, 28, 47, 57–59
Chandra X-ray Observatory, 74
charge, 8, 12–13
charm quarks, 9, 11, 83
closed universe, 62, 63
COBE. *See* Cosmic Background Explorer
color, galaxy movement and, 60
comets, 31, 32
compression, formation of matter and, 69
constellations, 33, 36
Copernicus, Nicolaus, 35–38, 41
Cosmic Background Explorer (COBE), 80
cosmic background radiation (CBR), 71–72, 75, 76–77
cosmic rays, 76
cosmological constant, 56
cosmological dimming, 61
cosmologists, defined, 32
cyclic model, 78–79
Cygnus X-1, 28, 65

D

daisy petal effect, 46, 53
dark energy, 20, 62, 77
dark matter, 16–19, 53, 77
deuterium, 9
Dialogue Concerning the Two Chief World Systems (Galileo), 40
Dicke, Robert, 72
dimming, cosmological, 61
distance, stellar, 57–58
Doppler, Christian, 58
Doppler effect, 56–59
down quarks, 9–10, 11

E

$E=mc^2$, 51
Eddington, Arthur Stanley, 53
Einstein, Albert
  black holes, singularities and, 65
  general theory of relativity of, 52–54, 56, 64, 87
  special theory of relativity of, 48–51
  time travel and, 52–55
Einstein rings, 18
Einstein's general theory of, 64
electromagnetic energy, 15, 60
electromagnetic force, 12–13, 67
electromagnetic radiation, 15, 16, 71–73
electromagnetic waves, 15, 60
electrons, 7–8, 11, 69
elementary particles, formation of, 69
elements, defined, 8
ellipses, 39
elliptical galaxies, 24
endless universe, 78–80
energy, matter and, 13–16
epicycles, 33, 35, 36
Eratosthenes, 33–34
eternal universe, 62–64
evolution, 78–79
exoplanets, 30

F

fate of universe, 62–64
flat universe, 63, 64
flatness problem, 76, 77
flavors, 9, 11, 13
Friedmann, Alexander, 56, 66
fundamental forces, overview of, 12–13
fundamental particles, 9–12, 84
fusion, nuclear, 25

G

galaxies, 22–24. *See also Specific galaxies*
Galileo Galilei, 39–43
gamma ray bursts, 16
gamma rays, 14, 15
Gamow, George, 66, 67, 71
Gemini North telescope, 30
general theory of relativity, 48–51, 56, 64, 87
geocentrism, 34, 35
gluons, 13
Gödel, Kurt, 54
Gold, Thomas, 72
Goldilocks zone, 75
Grand Unification Epoch, 67
Grand Unified Theory (GUT), 68–70, 86
gravitational lensing, 18, 51
gravity
  as fundamental force, 13
  general theory of relativity and, 51–52
  Grand Unification Epoch and, 67
  Newton's laws and, 44, 45–46
  slowing of universal expansion and, 61–63
Greenstein, Jesse, 24
GUT. *See* Grand Unified Theory
Guth, Alan, 67–69

H

Hadron Electron Ring Accelerator (HERA), 84
hadronic matter, 14
Hartle, Jim, 80
Hawking, Stephen, 42–43, 46, 76, 80
heat, as form of energy, 15
heavy hydrogen, 8
HERA. *See* Hadron Electron Ring Accelerator
helium, 26, 69, 75
heresy, 36, 40–41
Hertzsprung-Russell diagram, 25, 26
Higgs, Peter, 86
Higgs bosons, 86
horizon problem, 76–77
Hoyle, Fred, 67, 74
Hubble, Edwin, 58–61
Hubble Deep Field, 82
Hubble Space Telescope, 30, 61, 80–83

Hubble Ultra Deep
  Field, 82, 84
Hubble's law, 61
Hyades, 53
hydrogen atoms, 8, 26,
  69

I

imaginary time, 80
inflationary theory,
  67–69, 78
infrared (IR) light, 15
Inquisition, 41
instruments for
  astronomy, overview
  of, 79–83
iron formation, 73
isotopes, defined, 8

K

Keck Observatory, 30
Kepler, Johannes,
  39–40
Kepler's laws, 39
Kerr, Roy, 54
kinetic energy, 15

L

Large Hadron Collider
  (LHC), 85, 86
Large Magellenic Cloud
  galaxy, 26
Leavitt, Henrietta, 47,
  58–59
Lemaitre, Georges, 66
lensing, gravitational,
  18, 53
Lenticular galaxies, 24
leptons, 11, 13
LHC. See Large Hadron
  Collider
life, extraterrestrial, 75
light
  black holes and,
    64–65
  as form of energy, 15

  speed of, 58–53
  stellar distance and,
    57–58
light-years, 22
Lippershey, Hans, 39
Local Group, 22, 24
Local Supercluster, 24
Lou Gehrig's disease,
  42–43
luminosity, 57

M

M theory, 64
Mars, 36
Mather, John, 82
matter
  defined, 7
  energy and, 13–16
  formation of, Big
    Bang and, 69–71
  speed of, 48–51
mechanical energy, 15
Mercury, 44–45, 53
Michell, John, 27
microwaves, 15, 27,
  71–72
Milky Way galaxy,
  22–24, 28, 47–48
mirrors, telescopes and,
  43
moons, 31
muons, 85

N

nebulae, 28, 69
neutrinos, formation
  of, 69
neutron stars, 27
neutrons
  atoms and, 7–9
  formation of, 69
  quarks of, 10, 11
Newton, Isaac, 41–46,
  56
Newtonian telescopes,
  44
Nobel Prizes, 48, 72, 81

no-boundary universe,
  79, 80
North Star, 28
nuclear forces, Grand
  Unification Epoch
  and, 67
nuclear fusion, 25
nucleus, strong nuclear
  force and, 12–13

O

open universe, 62, 63
oscillating model,
  79–80

P

parallax, 57
particle accelerators,
  82–84
Penrose, Roger, 72–73
Penzias, Arno, 71–73
perihelions, 39
periodic table, 8
phase transitions,
  energy and, 15
photoelectric effect, 48
photons, 64–65, 69
pions, 14
planetary motion,
  Kepler's laws of, 39–40
planets, formation of,
  31
plasma, 69, 84
Polaris (North Star), 28
positrons, 14
potential energy, 15
protium, 8
protons
  atoms and, 7–8
  formation of, 69
  particle accelerators
    and, 83
  quarks of, 9, 11
Proxima Centauri, 22
Ptolemy (Claudius
  Ptolmaeus), 33–35

pulsars (pulsating stars), 27

Q

quark-gluon plasma, 84
quarks
  formation of, 69
  overview of, 8–9
  particle accelerators and, 82–84
  weak nuclear force and, 13
quasars, 24, 25

R

radiation, cosmic background, 71–73, 75, 76–77
radio waves, 15
redshift, 58–60
reflecting telescopes, 44
refracting telescopes, 44
Relativistic Heavy Ion Collider, 84
relativity
  Einstein's general theory of, 52–54, 56, 64, 87
  Einstein's special theory of, 48–51
retrograde motion, 34, 35–37
Rosen, Nathan, 54

S

Sanduleak-69° 202, 27–28
scale, of universe, 21–22
Schmidt, Maarten, 24
Schwarzchild, Karl, 64
scientific method, 38
Shapley, Harlow, 46–47
singularities, 62, 64–66, 68
Smoot, George, 82
smuons, 85

Solar System, overview of, 28–31
space probes, 81
space-time curvature, 50, 54
special theory of relativity, 58–61
spiral galaxies, 22–24
squarks, 85
standard model, 12
stars, 24–28, 38, 73. *See also Specific stars*
steady-state theory, 72–75
Steinhardt, Paul, 78–79
strange quarks, 9, 11
string theory, 64, 87
strong (nuclear) force, 13, 67
subatomic particles, overview of, 8–9
Sun
  as center of universe, 39–40
  formation of, 69
  Kepler's laws and, 39–40
  Milky Way galaxy and, 46–47
  overview of, 24–25
supercomputers, 84
supergiants, 28
supernovae, 16, 26–27, 73, 74
supersymmetry, 85
symmetry breaking, 67

T

telescopes
  detection of energy and, 15–16
  Gemini North telescope, 30
  Hubble Space Telescope, 30, 59, 80–82
  invention of, 39, 41, 44

overview of, 79–81
Theory of Everything, 87
time, space-time curvature and, 50
time travel, 54–55
top quarks, 9, 11
toroid magnets, 85
tritium, 8
Turok, Neil, 78–79
twin paradox, 50

U

Uhuru X-ray satellite, 28
ultraviolet (UV) light, 15
universe
  fate of, 62–64
  galaxies and, 22–24
  scale of, 21–22
  as singularity, 64, 66
  Solar System and, 28–31
  stars and, 24–28
up quarks, 9–10, 11
Ursa Major, 80–81

V

visible light, 15

W

wavelength, 15, 58, 61
waves, Doppler effect and, 56–59
weak (nuclear) force, 13, 67
weak bosons, 13
Wheeler, John Archibald, 27
whirlpools, 54
white dwarf stars, 25
Wilkinson Microwave Anisotropy Probe (WMAP), 82

Wilson, Robert, 71–73
WIMPs (weakly interacting massive particles), 19
W.M. Keck Observatory, 30

WMAP. *See* Wilkinson Microwave Anisotropy Probe
wormholes, 53–55

X

X-rays, 15, 28

Y

yellow dwarf stars, 25, 75

Z

Zodiac signs, 33
Zwicky, Fritz, 16–17, 19

About the Author

Kristi Lew is the author of more than 30 science books for teachers and young people. Fascinated with science from a young age, she studied biochemistry and genetics at North Carolina State University. Before she started writing full time, she worked in genetic laboratories for more than 10 years and taught high-school science. When she's not writing, she enjoys sailing with her husband aboard their small sailboat, *Proton*. She lives, writes, and sails in St. Petersburg, Florida.